MOTION AND TIME STUDY

Improving Work Methods and Management

Fred E. Meyers

Registered Professional Industrial Engineer
Associate Professor of Industrial Technology
Southern Illinois University

PRENTICE HALL
Englewood Cliffs, New Jersey 07632

Library of Congress Cataloging-in-Publication Data

Meyers, Fred E.
    Motion and time study : improving work methods and management /
Fred E. Meyers.
        p.   cm.
    Includes bibliographical references and index.
    ISBN 0-13-596081-9
    1. Motion study.   2. Time study.   3. Methods engineering.
I. Title.
T60.7.M48   1992
658.5′42—dc20   91-42372
                                                 CIP

Acquisition Editor: Rob Koehler
Production Editors: Fred Dahl and Rose Kernan
Copy Editor: Rose Kernan
Designers: Fred Dahl and Rose Kernan
Cover Designer: Ben Santora
Prepress Buyer: Ilene Levy
Manufacturing Buyer: Ed O'Dougherty
Supplements Editor: Judy Casillo

 © 1992 by Prentice-Hall, Inc.
A Simon & Schuster Company
Englewood Cliffs, New Jersey 07632

Printed in the United States of America
10 9 8 7 6 5 4 3 2 1

ISBN 0-13-596081-9

Prentice-Hall International (UK) Limited, London
Prentice-Hall of Australia Pty. Limited, Sydney
Prentice-Hall Canada Inc., Toronto
Prentice-Hall Hispanoamericana, S.A., Mexico
Prentice-Hall of India Private Limited, New Delhi
Prentice-Hall of Japan, Inc., Tokyo
Simon & Schuster Asia Pte. Ltd., Singapore
Editora Prentice-Hall do Brasil, Ltda., Rio de Janeiro

Dedicated to my wife, Mary, and our children,
Fred, Vicki, Jeff, and Mike

Contents

CHAPTER 7
Work Station Design, 75

CHAPTER 8
Predetermined Time Standard System, 90

CHAPTER 9
Stopwatch Time Study, 114

CHAPTER 10
Standard Data, 166

Preface

The purpose of this how-to motion and time study book is to provide students and practitioners with a resource which describes the techniques and procedures of motion and time study. This book has appropriately been called a "cookbook." Practial, detailed advice is given on all aspects of motion and time study including work station design, job analysis, and the techniques of setting time standards.

The mathematics requirement of this textbook is high school algebra. A few simple formulas are included in the standard data chapter. These formulas require the insertion of a variable to calculate the time requirement. Two more complicated formulas are used to show how tables are developed. The practitioner would use the tables to save time.

Manufacturing plant management needs time standards. Many major decisions would be only a guess without time standards. How would we determine how many machines to buy, how many people to hire, how much to sell the product for; how would we schedule the plant, how would we justify new methods or equipment, how would we ensure a balanced work load on assembly lines, and how would we evaluate employees or pay for increased effort? Chapter 3 answers these questions and inspires an appreciation of the importance of motion and time study.

Motion study is accomplished before time standards are set. When a company decides to introduce a new product, a technician will be asked to provide a plan to produce, for example, 1,500 units per day. The technician must design work stations for every fabrication, assembly, and packout operation. From the work station drawing, a left-hand/right-hand analysis of the work content is made. A predetermined time standard has been set for every body motion, so the time for every motion required to do the job is added together. This will be the time standard, and it was set before the company had the first part, a machine, or operator. Chapters 5 through 8 discuss methods analysis techniques.

Stopwatch time study can be accomplished only after the machines have been installed and the operator fully trained. In a proposed new plant, no machine and people are available to time study, but an operating plant could use stopwatch time study very effectively. The stopwatch technique is the oldest technique of setting time standards and it is entrenched in many companies. Chapter 9 examines this technique.

Standard data is another technique of setting time standards before production begins, but it is developed from in-plant experience. Standard data is very personal to a specific company, and companies cannot normally use another's standard data. This is the most accurate, least costly method of setting time standards, and every motion and time study department should be developing theirs. Chapter 10 examines this technique.

Work sampling is based on the laws of probability and is a scientific technique of setting quality time standards. Office work, engineering departments, maintenance craft, and even equipment can be work sampled. Everyone who has worked with others has work sampled. Chapter 11 discusses this common practice in a scientific light. Consultants often use work sampling first to establish the beginning efficiency of the operation. Potential savings forecasts will be based on current efficiency.

Labor is a significant portion of manufacturing cost and must be controlled. Performance control systems based on time standards give management the control they need. History and research have shown that operations working without a performance control system average 60% of normal. When a performance control system is established, 85% performance results. Industrial plants on incentive average 120% performance. The size of these cost reductions is spectacular, and no industrial technologist or industrial manager will go unnoticed when such improvements are made. Chapter 12 discusses performance control systems.

Chapters 13 and 15 are uses of time standards that could be an important part of a technologists career. Wage payment, chapter 13, includes incentive systems, which are a fun area for technologists to work. The assembly line balancing, chapter 15, is instructions on setting up assembly lines. This is a big area in many plants.

Chapter 14 is a military standard. If you want to provide product for the government, you must abide by their rules. Chapter 14 deals with the motion and time study rules required by the federal government.

Chapter 16, the Time Management techniques chapter is aimed at making the motion and time study technologist more productive.

Human relationships are an important part of motion and time study. The successful attitudes and goals of motion and time study technologists are discussed in chapters 17 and 18.

A step-by-step procedure, real-life examples, sample problems, and blank forms are included for every technique. This book will remain a good reference source years after a course or seminar on motion and time study.

Your feedback will be valued and considered for future editions. Our objective is to provide a practical, usable, how-to text for motion and time study.

About the Author

Fred E. Meyers is a Registered Professional Industrial Engineer, Associate Professor of Industrial Technology, Director of the College of Engineering and Technology Applied Research Center, and the 1988–1989 Outstanding Teacher of the College of Engineering and Technology at Southern Illinois University at Carbondale, Illinois. He has fourteen years of industrial engineering and production management experience with such companies as Caterpillar Tractor Company, Mattel Toy Company, Boeing Aerospace division, Ingersol-Rand's proto tool division, and Spalding Golf Club division. Professor Meyers has worked as an industrial consultant since joining Southern Illinois University in 1975. He has consulted with over fifty companies in many different industries such as energy, oil, sporting goods, transportation, appliances, distribution, lumber and plywood, paper manufacturing, furniture, tooling, fiberglass, and office work. His variety of assignments has given him the ability to see the wide-ranging uses of motion and time study.

Fred E. Meyers has taught motion and time study to over 100 classes and 4,000 students. He has presented seminars to the National Association of Industrial Technology, industrial plants, and labor unions.

Acknowledgments

I am a student of Ralph Barnes, Peter August, and Mitchell Fein. They have influenced me greatly, and their attitudes are a part of me. I must also thank Dr. Matthew P. Stephens for his statistical expertise, Dr. Dale H. Besterfield for his example, Dr. James P. Orr for his encouragement, and the faculty of the Department of Technology for giving me the time to write this book.

Introduction to Motion and Time Study

Technology is the application of techniques. Few techniques are new, and no technique is ever truly obsolete. The study of motion and time study is purely the study of technique. Our production equipment, work methods, and manpower are getting better every year, but the need for application of motion and time study techniques will never be obsolete.

A technologist is one who studies technique. Industrial engineering technology and industrial technology students are being prepared to design work stations, develop better work methods, establish time standards, balance assembly lines, estimate labor costs, develop effective tooling, select proper equipment, lay out plants, train workers in effective and efficient methods, and train other managers to be methods conscious. This is what industry is looking for.

This book is divided into three sections:

I. Methods analysis techniques

II. Time-standard–setting techniques

III. Uses of time standards.

An industrial technologist working with motion and time study will study an industrial job or series of jobs by learning the details of that job and making changes. Changes may be small but must be continual. Industry can never stop looking for better methods, or it will become obsolete. The industrial technologist becomes a part of the system and the motivator of improvement. Time standards are a way of keeping score. Time is money, and a time standard tells us exactly how much money.

Few industries have new technology that is exclusively theirs. What industry has that is more important than exclusive technology is a technologist who understands that improvement comes only by hard work and attention to detail. There is no easy way.

Breaking down a job into its smallest components and putting it back together again using motion study techniques will result in an improvement. A motion and time study technologist will have the following attitudes:

"We can reduce the cost of any job."

"Cost is our measuring rod."

"Cost reduction is our job."

American industry must continue to deliver quality products at a reasonable price. Quality and price are the most important considerations for staying competitive. Motion and time study technologists concentrate on reducing costs but must never lose sight of quality. The following attitudes are critical:

"We never propose a method that will reduce quality."

"We never set standards for producing scrap."

"Lower cost and high quality are our competitive edge. One without the other leads to failure."

"Work smarter, not harder" should be the motto of every industrial technologist. The techniques of motion and time study can reduce the cost of any job, and the savings can be spectacular.

Motion study has the greatest potential for savings. We can save nearly the total cost by eliminating the task or combining the task with some other task. We can rearrange the elements of work to reduce the work content, and we can simplify the operation by moving parts and tools closer to the point of use or downgrading the complexity of motions by using tools, fixtures, or redesigning parts of the product. Simplification is the most time-consuming way of reducing costs, and its savings are small compared to elimination or combining—but we can always simplify.

Time study can reduce cost significantly as well. Time standards are goals to strive for. In organizations that operate without time standards, 60% performance is typical. This statistic can be proven by work sampling that operation. When time standards are set, performance improves to an average of 85%. This is a 42% increase in performance:

$$\frac{85\% - 60\%}{60\%} = 42\% \text{ productivity increase.}$$

Incentive systems can improve performance even further. Incentive system performances average 120%, another 42% increase in productivity:

$$\frac{120\% - 85\%}{85\%} = 42\% \text{ productivity increase.}$$

1. Manufacturing plants with no standards average 60% performance.

2. Manufacturing plants with time standards average 85% performance.

3. Manufacturing plants with incentive systems average 120% performance.

If more production is required, don't buy more machinery, don't add a second shift, don't build a new plant: Just establish a motion and time study program.

Motion and time study is a lot of work and creates some labor/management conflict, but the need is so important that it must be done. It is said that successful people do what other people do not want to do:

1. Work long.

2. Work hard.

3. Criticize.

4. Be criticized.

Even though our motto should be "work smarter not harder" we need to continue to work hard. Motion and time study will fit this definition.

Motion and time study is considered to be the backbone of the industrial technology and industrial management programs because the information such study generates affects other areas, such as

1. Cost estimating;

2. Production and inventory control;

3. Plant layout;

4. Materials and processes.

5. Quality.

6. Safety.

Motion and time study creates a cost consciousness that is desired in every manufacturing manager or engineer. Cost conscious engineers and managers have a competitive advantage on all other engineers and managers. A description of the uses of time study follows in chapter 3.

Motion study precedes setting time standards. An industrial technologist's time would be wasted setting time standards for poorly designed jobs. Cost reduction via motion study is automatic and can be significant. Motion study is a detailed analysis of the work method to improve it. Motion studies are used to

1. Develop the best work method.

2. Develop motion consciousness on the part of all employees.

3. Develop economical and efficient tools, fixtures, and production aids.

4. Assist in the selection of new machines and equipment.

5. Train new employees in the preferred method.

6. Reduce effort and cost.

Motion study is for cost reduction, and time study is for cost control. Motion study is the creative activity of motion and time study. Motion study is design, while time study is measurement.

This book focuses on techniques. Once the importance of motion and time study is understood and accepted, the techniques of motion and time study are introduced. Technique is the *how* of motion and time study.

The following techniques of motion study are covered in this book:

1. Process charts
2. Flow diagrams
3. Multiactivity charts
4. Operation charts
5. Flow process charts
6. Predetermined time standards system.

The techniques of time study start with the last motion technique, and this shows the close relationship between motion study and time study. The techniques of time study are as follows:

1. Predetermined time standards system
2. Stopwatch time study
3. Standard data time standards
4. Work sampling time standards
5. Expert opinion standards.

Each of these techniques is covered in chapter 3 and again in great detail in a later chapter. *How-to* is the core of this book. This is a techniques book. Each technique includes a step-by-step procedure for using this industrial technology tool. Each technique also includes a completed example and a problem to solve. At the back of the book are blank forms to be used and reproduced as needed. You have the author's permission to copy these forms for your or your company's use. The heading of the forms can be easily changed to your company name.

QUESTIONS

1. What is technology?
2. What does an industrial technologist do?
3. What are the two most important considerations for a company to stay competitive?

4. What are three good attitudes of an industrial technologist?

5. What should be the motto of industrial technology?

6. What are four ways motion study can reduce cost? What is the most important way?

7. At what percentage of performance do manufacturing plants operate in the following situations:
   a. No time standards? _____
   b. With time standards? _____
   c. With incentive standards? _____

8. What are the four things successful people do that unsuccessful people do not want to do?

MOTION AND TIME STUDY

Semester Project

OBJECTIVE

(Teams of Three or Four People)

Provide the student with real experience in motion and time study. Each student or team of students will design, build, and operate a work station to assemble a simple product.

MINIMUM REQUIREMENTS

1. Time constraint
   The job must take at least 15 seconds (.25 minutes). If you do two at a time, 15 seconds for two would be OK.

2. Operation of the work station
   Operate your work station for 45 minutes while class members time-study you.

3. Written report must include
   a. Product drawing
   b. Work station drawing and layout of parts
   c. Motion pattern analysis
   d. Predetermined time standard (PTSS)
   e. Write-up of the principles of motion economy used in your work station.

4. You will also be graded on
   a. Accuracy of PTSS and time study
   b. Professionalism
   c. Operation of work station.

CHAPTER 2

History of Motion and Time Study

The history of motion and time study is not long but has been full of controversy. Time study was developed in about 1880. Frederick W. Taylor is said to be the first user of the stopwatch to measure work content. His purpose was to define "a fair day's work." In about 1900, Frank and Lillian Gilbreth started working with methods study. Their goal was to find the one best method. During 1928, Elton Mayo started what is known as the human relations movement. By accident, he discovered that people work better when their attitude is better. These four pioneers of motion and time study are discussed in this chapter. But first, we provide a little earlier background.

Labor has always been a major factor in the cost of a product. As labor productivity improves, costs go down, wages go up, and profits go up. From the earliest industrial history, management has looked for labor-saving techniques. Industrial technology's objective and purpose for being is to increase productivity and quality. Output per manhour is the most commonly used measurement of productivity. Motion and time study techniques give management the tools to measure and improve productivity.

Concern for productivity has always been a major motivation of production managers. Productivity is a concern of anyone in business. Look at the farmer, for example. How much more work can be accomplished with a tractor over horses? How many more acres can be plowed, planted, and reaped by a single person? How many more bushels per manhour can be harvested? Bushels per manhour is a good measure of farm productivity. Take this concept into manufacturing, and we have number of units produced per manhour. Coal mining is another good example. Everyone agrees that mining machines are more productive than pick and shovel. Would anyone disagree that the mining job of today is better than 100 years ago? Of course not! The tons of coal per laborer per day has continued to improve.

Historical examples of how new technology increased productivity in every area of business and industry exist by the thousands. The steam engine replaced horsepower, making the Industrial Revolution possible, and interchangeable parts replaced one-of-a-kind parts, making mass production and the assembly line possible. During the nineteenth century, early manufacturers were vying for competitive advantages, and technology was expanding at a fever pitch. Secrecy was of utmost importance. Sharing of information and ideas was minimal.

In 1832, Charles Babbage wrote his book *On the Economy of Machinery and Manufacturers* and included ideas on division of labor, organization charts, and labor relations. It would be 50 to 100 years before they were used extensively. Higher education and professional societies are given credit for opening up the information on manufacturing management techniques, but the sharing of information was slow because of the secrecy mentality of most managers. Only in Frederick W. Taylor's later life did he write about what he had done.

FREDERICK W. TAYLOR (1856–1915)

Frederick W. Taylor is known as the father of scientific management and industrial engineering. He is the first person to use a stopwatch to study work content and, as such, is called the Father of Time Study.

He was born in Philadelphia, Pennsylvania, to wealthy parents. He passed the Harvard University entrance exams with honors, but eye problems stopped him from attending. Taking his doctor's advice, Taylor entered the labor force as an apprentice machinist. Four years later, at the age of 22 (1878), Taylor went to work for Midvale Steel Works as a laborer. He worked his way up to time keeper, journeyman, lathe operator, gang boss, and foreman of the machine shop. At age thirty-one, he was Chief Engineer of Midvale Steel Works. In 1883, after years of night school, he earned a B.S. in Mechanical Engineering from Stevens Institute.

Many years later, Taylor was able to explain what he accomplished through his four Principles of Scientific Management:

1. Develop a science for each element of a person's work, thereby replacing the old rule-of-thumb methods.
2. Select the best worker for each task and train that worker in the prescribed method developed in Principle 1.
3. Develop a spirit of cooperation between management and labor in carrying out the prescribed methods.
4. Divide the work into almost equal shares between management and labor, each doing what they do best.

Before Taylor, the work force developed its own methods through trial and error. The work force was responsible for seeing that everything was available to do the job, such as laborers bringing their own tools to work.

Frederick Taylor wanted management to reject opinion for a more exact science.

Taylor:

1. Specified the work method;

2. Instructed the operator in that method;

3. Maintained standard conditions for performing that work;

4. Set time standard goals;

5. Paid premiums for doing the task as specified.

    Frederick Taylor is responsible for the following innovations:

1. Stopwatch time study

2. High-speed steel tool

3. Tool grinders

4. Slide rules

5. Functional type organization.

Taylor's Shoveling Experiment

Between 400 and 600 men moved mountains of coal, coke, and iron ore around the 2-mile-long yards of Midvale Steel Works. Each man brought his own shovel from home and was assigned to a gang for moving materials. Taylor noticed the different sizes of shovels and wondered which shovel was the most efficient. Taylor talked management into a formal study of this operation. He asked a laborer, known today only as John, if he would be willing to help him study the coal, coke, and iron ore shoveling job. Taylor told John he would double his salary, and I'll bet John answered positively within 10 seconds. Taylor watched John with a stopwatch and measured everything he did. He varied the shovel size, duration, number of breaks, and work hours. The results were fantastic. He purchased quantities of three types of shovels—one for coal, one for coke, and one for iron ore.

TABLE 2-1

	Before Study	After Study
No. people	400–600	140
Pounds/shovel	$3\frac{1}{2}$–38	$21\frac{1}{2}$
Bonus	No	Yes
Work unit	Teams	Individual
Cost/ton	7¢ to 8¢	3¢ to 4¢

A savings of $78,000/year.

FRANK (1868–1924) AND LILLIAN (1878–1972) GILBRETH

Frank and Lillian Gilbreth are known as the parents of motion study. In their lifetime search for the one best method of doing a specific job, they developed many new techniques for studying work. Their title as parents of motion study is universally accepted.

Frank Gilbreth started his work life as a bricklayer's apprentice. He was immediately aware of motions. He noticed that when the instructor showed him how to lay bricks he used one set of motions, another set of motions when he was working by himself, and a third set of motions when in a hurry. Frank questioned this practice and went about searching for the one best method. He set up his own construction business with the competitive advantages of

1. Adjustable scaffolding—previously bricklayers built the wall from their toes to their highest reach, then built some more scaffolding and started over.
2. Bricklayer's helpers—at about one half the cost of a bricklayer. The helper would sort, carry, and stack bricks for the bricklayer.
3. Constant mortar mix.
4. Improved motion pattern.
5. Three hundred and fifty bricks per hour instead of the previous 120.

Lillian Gilbreth (shown in Figure 2-1) was a trained psychologist and a people-oriented person. In addition to her work in methods engineering, she raised twelve

FIGURE 2-1 An outstanding pioneer of motion study

children and authored books, one of which was *Cheaper by the Dozen*. I know that Lillian's response to Frank when he came home from work saying, "You should have seen how I designed that job today," was "Frank, you can't do that to people." Lillian kept Frank from dehumanizing work and made him conscious of the human element.

Frank and Lillian's continued success in motion study led them away from the construction business into consulting. Frank's training in engineering and Lillian's training in psychology made them a powerful team. Not only did they develop methods study techniques like the cyclograph, (shown in Figure 2-2), chronocyclographs, movie cameras, etc., they also studied fatigue, monotony, transfer of skills, and assisted the handicapped in becoming more mobile. Their work now is a tradition of industrial engineering.

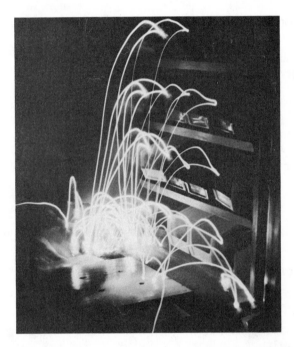

FIGURE 2-2 Gilbreths' cyclograph: The light path the hands make in the process of producing one part or cycle.

The knack which Frank and Lillian possessed for analyzing work motion situations enhanced their ability for substituting shorter and/or less fatiguing motions to improve the work environment. Their research has called attention to the fact that great gains may be realized, even in the simpler trades where one would not suppose such possibilities exist. Their systematic study of motions reduced costs greatly and founded a new profession of methods analysis.

The elimination of all useless motions and the reduction of those remaining motions was the foundation of the Gilbreths' work. The elimination of this unwanted waste has become known as *work simplification*. The Gilbreths used flow diagrams to show movement of product around an entire plant because they gave an accurate geographical picture of the entire process. The Gilbreths developed process charts to show diagramatically the sequence and relationship of elements of a process. The operations chart showed detail of the individual operations. These charts were able to show the interrelationship of laborer and machine, gangs of people working together, and left hand versus right hand. Some motions were so fast that the Gilbreths incorporated a high-speed motion picture camera and a special clock called a microchronometer to study this work. Increments of 1/2000 of a minute were used. From this work came micro-motion study.

Among the generally accepted theories of efficient motions developed by the Gilbreths was the terminology defining the entire range of manual motions. To refer to these seventeen elementary subdivisions of motion, later engineers coined a short word, *therblig,* which is Gilbreth spelled backwards (except for the *th*). Therbligs are the basic components of the motion pattern. Essentially, they name the different activities of the hand. Each therblig has a symbol, color designation, and letter symbol for the purpose of charting. Search, select, grasp, transport, hold, position, inspect, and assemble are a few of the therbligs.

The predetermined time standard system that is studied in this book is a newer, easier to learn, easier to apply technique built on the Gilbreths' work.

PROFESSOR ELTON MAYO

Known as the father of the human relations movement, Professor Elton Mayo became involved in the Western Electric Company's Hawthorne plant productivity studies after the National Research Council of the National Academy of Science withdrew.

The Hawthrone plant, near Chicago, undertook a research project to study what factors affected productivity. These studies took place between 1924 and 1933.

Phase I (1924–1927): Illumination Study

The basic premise of this study was that increased illumination of the work area would increase productivity. The results of this first three-year study were called inconclusive because too many other factors entered the picture. The National Research Council withdrew from the study, but Professor Elton Mayo of Harvard University became interested in Western Electric's efforts and joined forces for the next phase.

Phase II (1927–1929): Relay Assembly Study

A group of five women were selected, briefed, and relocated to an experimental assembly room where other factors could be controlled (other factors were blamed for the failure of the illumination study). The basic premise of the relay assembly experiments was that "a change in working conditions would result in a change in productivity."

To isolate the factors being studied, the experimenters worked hard to keep the operators' attitudes positive toward the study, management, and their work. The experiment observers spent much time talking to and listening to the operators.

The factors being studied were

1. Incentive system
2. Rest periods
3. Paid lunch breaks
4. Elimination of Saturday work
5. Reduction in work hours
6. Free lunch and drinks.

The relay assembly experiments consisted of thirteen phases. Table 2-2 shows the phases, factors, duration, and results.

Phase III (1929–1930): Interviewing Program

Twenty-one thousand employees of Western Electric Corporation were interviewed. The company wanted their opinions on what people want from their jobs. Learning how to ask questions and learning how to listen were the greatest contributions of this phase.

TABLE 2-2 HAWTHORNE PLANT: RELAY ASSEMBLY EXPERIMENT THIRTEEN PHASES

Phase	No. Weeks Duration	Cumulative No. Weeks	Factor	Hours Worked per Person	Average Hourly Output	Average Weekly Output
1	2	2	In regular dept.	39.7	49.51	1,973
2	5	7	Experiment room	45.6	49.4	2,254
3	8	15	Incentive system	44.7	51.25	2,289
4	5	20	2–5 min. rest	42.4	53.11	2,251
5	4	24	2–10 min. rest	44.2	55.9	2,470
6	4	28	6–8 min. rest	44.2	55.9	2,470
7	11	39	11 min A.M.– Lunch—10 min. rest	41.4	56.1	2,305
8	7	46	Stop $\frac{1}{2}$ hr. early	40.7	62.5	2,542
9	4	50	Stop 1 hr. early	39.0	64.5	2,516
10	1	62	Same as 7	43.6	61.7	2,691
11	9	71	No. 7 + eliminate Sat. A.M.	39.6	63.6	2,517
12	12	83	No. 3 no lunch/rests	45.9	61.0	2,802
13	31	114	Same as no. 7 but brought own lunch— given beverages	43.1	66.7	2,873

Phase IV (1931–1932): Bank Wire Observation Room

Informal organization and its influence on productivity were studied in this phase.

The results of the Hawthorne studies did not go as intended. The factors thought to improve performance did not lead to an automatic improvement. The studies did show that change affects the employee's attitude, which in turn affects results. The experimenters treated the operators differently than was typical of the period. As a result, the operators were made to feel important, and they were involved with something they thought important. When the other factors were changed negatively, production still improved, because the employees' attitudes continued to be positive. Even the early studies of illumination became proof of this new hypothesis, "Improve employee attitude and you will improve productivity." In modern times, company management is involving the work force in all phases of product development and implementation. Modern companies are finding the whole person to be of great help in all areas that affect them. No more do we consider people hired hands. We get the whole thing—hands, mind, mouth. It may have been easier in the old days, but bringing all minds to bear on a problem gives us better solutions.

OTHER PIONEERS

Many other people have been involved with the development of motion and time study. Industrial engineers and technologists continue to improve techniques every day, but time prevents us from mentioning all the pioneers.

CONTROVERSY

We mentioned controversy about motion and time study in chapter 1. We also mentioned that successful people do what others do not like to do, among which are to criticize and be criticized.

Frederick W. Taylor was criticized as being a management speed-up artist. Unscrupulous managers used Taylor's techniques, and when workers met the goals, management raised the standard. Taylor would hate this process, and we must never change a standard without due cause.

Lillian and Frank Gilbreth were charged with dehumanizing work. Because of the reduction of motions to the absolute best set of procedures possible, unions depicted the Gilbreths as antiworker and as wanting to make machines out of everyone. The Gilbreths are not properly given credit for removing the drudgery from work.

Elton Mayo was charged with being unscientific. His improvements were said to be caused by other factors.

Anytime anyone does something new and different, people criticize. Criticism is easier than developing something new, so many would-be scientists try to show that some past study is in error.

Frederick Taylor, Frank and Lillian Gilbreth, and Elton Mayo were hard-working, ethical human beings who improved our world of work and had the courage to write about it.

QUESTION

Identify the contributions made to the field of motion and time study by the four pioneers discussed in this chapter.

CHAPTER 3

The Importance and Uses of Motion and Time Study

The time standard òf one of the most important pieces of information produced in the manufacturing department. Time standards are used to

1. Determine the number of machines or tools required.
2. Determine the number of production people to hire.
3. Determine manufacturing costs and selling price.
4. Schedule machines, operations, and people to get the work done on time.
5. Determine assembly line balance and plant layout of all departments and equipment.
6. Determine individual work performance and identify operations that are having problems.
7. Pay incentive wages for improved and outstanding performance.
8. Evaluate cost reductions and choose the best method.
9. Evaluate new equipment purchases and justify their expense.
10. Develop operation manpower budgets to measure management performance.

WHAT IS A TIME STANDARD?

A time standard is the time required to produce a product at a work station with the following conditions:

1. A qualified, well-trained operator;
2. Working at a normal pace; and
3. Doing a specific task.

A Qualified, Well-Trained Operator is required. Experience is usually what makes a qualified, well-trained operator, and time on the job is our best indication of experience. The time required to become qualified varies with the job and the person. For example, sewing machine operators, welders, upholsterers, machinists, and many other high-technology jobs require long learning periods. The greatest mistake made by new time study personnel is time-studying someone too soon. A good rule of thumb is to start with a qualified, fully trained person and to give that person two weeks on the job prior to the time study. On new jobs or tasks, predetermined time study systems are used. These standards seem tight (hard to achieve) at first because the times are set for qualified, well-trained operators.

Normal Pace is a concept we spend much time with in chapter 9. Only one time standard can be used for each job, even though individual differences of operators cause different results. A normal pace is comfortable for most people. In the development of the normal pace concept, 100% will be the normal pace. Time standards of normal pace commonly used are

1. Walking 264 feet in 1.000 minutes (3 miles per hour);
2. Dealing fifty-two cards into four equal stacks in .500 minutes;
3. Filling a thirty-pin pinboard in .435 minutes.

Training films for rating are also used to develop this concept.

A Specific Task is a detailed description of what must be accomplished. The description of the task must include

1. The prescribed work method
2. The material specification
3. The tools and equipment being used
4. The positions of incoming and outgoing material
5. Additional requirements like safety, quality, housekeeping, and maintenance tasks.

The time standard is only good for this one set of specific conditions. If anything changes, the time standard must change.

The written description of a time standard is important, but the mathematics is even more important. If a job takes 1.000 standard minutes to produce, we can produce sixty pieces per hour, and it will take .01667 hours to make one unit or 16.67 hours to make 1,000 units. The time in decimal minutes is always used in time study because the math is easier. The following three numbers are required to communicate a time standard:

1. The decimal minute (always in three demical places, e.g., .001)
2. Pieces per hour (rounded off to whole numbers, unless less than ten per hour)

Time Standard Minutes	Pieces per Hour[a]	Hours per Piece[b]	Hours per 1,000 Pieces[c]
1.000	60	.01667	16.67
.500	120	.00833	8.33
.167	359	.00279	2.79
2.500	24	.04167	41.67
.650			
.050			

[a]Pieces per hour is calculated by dividing the time standard minutes into sixty minutes per hour.

[b]Hours per piece is calculated by dividing the pieces per hour into one hour.

[c]Hours per 1,000 pieces is calculated by multiplying the hours per piece by 1,000 pieces.

3. Hours per piece (always in five decimal places, e.g., .00001). Many companies use hours per 1,000 pieces because the numbers are more understandable or meaningful.

Now that we understand what a time standard is, let's look at why time standards are considered to be one of the most important pieces of information produced in the manufacturing department.

USES OF TIME STANDARDS

How would you answer the following questions?

1. How many machines do we need?

The first question raised when setting up a new operation or starting production on a new product is "How many machines do we need?" The answer depends on two pieces of information:

a. How many pieces do we need to manufacture per shift?

b. How much time does it take to make one part? (This is the time standard.)

Example:

1. The marketing department wants us to make 2,000 wagons per eight hour shift.
2. It takes us .400 minutes to form the wagon body on a press.
3. 480 minutes per shift (8-hour shift)
4. −50 minutes downtime per shift (breaks, clean-up, etc.)

5. 430 minutes per shift available @ 100%

6. @ 75% performance (based on history or expectation) (.75 × 430 = 322.5)

7. 322.5 effective minutes left to produce 2,000 units

8. $\dfrac{322.5}{2{,}000 \text{ units}}$ = .161 minutes per unit or 6.21 parts per min.

This is called the plant rate. Every operation in the plant must produce a part every .161 minutes; therefore how many machines do we need for this operation?

$$\frac{\text{Time standard} = .400 \text{ minutes/unit}}{\text{Plant rate} = .161 \text{ minutes/unit}} = 2.48 \text{ machines}$$

2.48 machines are needed for this operation. If other operations are required for this kind of machine, we would add all the machine requirements together and round up to the next whole number. In the preceding example, we would buy three machines. (Never round down on your own. You will be building a bottleneck in your plant.)

2. How Many People Should We Hire?

Look at the operations chart shown in Figure 3-1. From a study of this chart, we find the time standard for every operation required to fabricate each part of the product and each assembly operation required to assemble and pack the finished product.

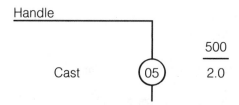

In this operation (casting the handle), the 05 indicates the operation number. Usually, 05 is the first operation of each part. The 500 is the pieces per hour standard. This operator should produce 500 pieces per hour. The 2.0 is the hours required to produce 1,000 pieces. At 500 pieces per hour, it would take us two hours to make 1,000. How many people would be required to cast 2,000 handles per shift?

$$\begin{array}{r} 2{,}000 \text{ units} \\ \times\, 2.0 \text{ hours/1,000} \\ \hline 4.0 \text{ hours at standard} \end{array}$$

Not many people, departments, or plants work at 100% performance. How many hours would be required if we work at the rate of 60%, 85%, or 120%?

$$\frac{4 \text{ hours}}{60\%} = 6.66 \text{ hours}; \quad \frac{4 \text{ hours}}{85\%} = 4.7 \text{ hours}; \quad \frac{4 \text{ hours}}{120\%} = 3.33 \text{ hours}$$

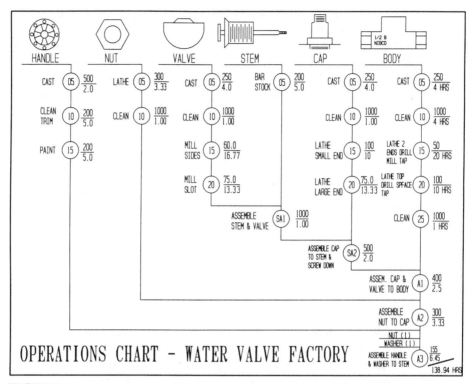

FIGURE 3-1 *Operations Chart* for a Water Valve Factory: A circle for every fabrication, assembly, and packout operation.

Therefore, depending on anticipated performance, we will be budgeted for a specific number of hours. Either performance history or national averages would be used to factor the 100% hours to make them practical and realistic.

Look again at the operations chart shown in Figure 3-1. Note the total 138.94 hours at the bottom right side. The operations chart includes every operation required to fabricate, paint, inspect, assemble, and pack out a product. The total hours is the total time required to make 1,000 finished products. In our water valve factory, we need 138.94 hours at 100% to produce 1,000 water valves. If this is a new product, we could expect 75% performance during the first year of production. Therefore,

$$\frac{138.94 \text{ hours per } 1,000}{75\% \text{ performance}} = 185 \text{ hours}/1,000$$

where 75% = .75.

The marketing department has forecasted sales of 2,500 water valves per day. How many people are needed to make water valves?

$$185 \text{ hours}/1,000 \times 2.5 \ (1,000) = 463 \text{ hours/day needed}$$

Divide this by 8 hours per employee per day, equal to fifty-eight people.

Management will be judged by how well it performs to this goal. If less than 2,500 units are produced per day with the fifty-eight people, management will be over budget, and that is unforgivable. If it produces more than 2,500 units per day, management is judged as being good at managing, and the managers are promotable.

Most companies produce more than one product. The problem of how many people to hire is the same. For example, How many direct labor employees do we need for a multiproduct plant?

Product	Hours/1,000	No. of Units Needed/Day	Hours at 100%	Actual %	Actual Hours Needed
A	150	1,000	150.0	70	214
B	95	1,500	142.5	85	168
C	450	2,000	900.0	120	750
					Total 1,132 hours

Per day, 1,132 hours of direct labor are needed. Each employee will work 8 hours; therefore,

$$\frac{1,132 \text{ hours}}{8 \text{ hours/employee}} = 141.5 \text{ employees.}$$

We will hire 142 employees; we will budget 142 employees. Our management will be evaluated on the performance of these 142 employees. Without time standards, how many employees would you hire? Any other method would be a guess. The need for high-quality time standards is illustrated by this example. Management doesn't want to be judged and compared to unattainable time standards or production goals.

3. How Much Will Our Product Cost?

At the earliest point in a new product development project, the anticipated cost must be determined. A feasibility study will show top management the profitability of a new venture. Without proper, accurate costs, the profitability calculations would be nothing but a guess.

Product costs consist of

		Typical %
Manufacturing costs 50%	Direct labor	8
	Direct materials	25
	Overhead costs	17
	Plus	___
Front-end costs 50%	Sales & distribution costs	15
	Advertising	5
	Administative overhead	20
	Engineering	3
	Profit	7
		100%

Direct labor cost is the most difficult component of product cost to estimate. Time standards must be set prior to any equipment purchase or material availability. Time standards are set using predetermined time standards or standard data from blueprints and work station sketches. The time standards are collected on something like the operations chart shown in Figure 3-1. On the bottom right side of the water valve operation sheet, we found the hours required to produce 1,000 units:

$$\frac{138.94 \text{ hours per } 1,000 \text{ units}}{85\% \text{ anticipated performance}} = 163.46 \text{ hours}/1,000$$

$$\begin{array}{ll} 163.46 & \text{hours per } 1,000 \text{ water valves} \\ \times \$ \; 7.50 & \text{per hour labor rate} \\ \hline \end{array}$$
$$\$1,225.94/1,000 \text{ or } \$1.23 \text{ each}$$

Direct material is the material that makes up the finished product and is estimated by calling vendors for a bid price. Normally, 50% of the manufacturing cost (direct labor + direct materials + factory overhead) is direct material cost. For our example, we will use 50%. On the operating chart, raw materials are introduced at the top of each line. Buyout parts are introduced at the assembly and packout station.

Factory overhead costs are all expenses of running a factory, except the previously discussed direct labor and direct material. Factory overhead is calculated as a percentage of direct labor. This percentage is calculated using last year's actual costs. All manufacturing costs for last year are divided into three classifications:

Direct labor	$1,000,000
Direct material	3,000,000
Overhead	2,000,000
Total factory costs	$6,000,000

The factory overhead rate for last year is

$$\frac{\$2,000,000 \text{ overhead cost}}{\$1,000,000 \text{ direct labor costs}} = 200\% \text{ overhead rate/labor dollar.}$$

Thus, each dollar in direct labor cost has a factory overhead cost of $2.00.

Example:	Water valve		
	Labor	$1.23	from time standards
	Overhead	$2.46	200% overhead rate
	Material	$3.69	from our suppliers
Total factory cost		$7.38	
All other costs		7.38	from ratios
Selling price		$14.76	

Labor cost is the most difficult cost to calculate of all the costs that make up selling price. How could you calculate selling price without time standards? Anything else is a guess.

Cost estimating is an important part of any industrial technology program and should be a complete course covering operations, product, and project costing. Motion and time study would, of course, be a prerequisite.

4. When should we start a job, and how much work can we handle with the equipment and people we have? Or, how do we schedule and load machines, work centers, departments, and plants?

Even the simplest manufacturing plant must know when to start an operation for the parts to be available on the assembly line. The more operations, the most complicated the scheduling.

Example: One machine plant operates at 90%.

Backlog Job	Hours/1,000	Units Required	Hours Required	Backlog—Cumulative Hours	Backlog—days
A	5	5,000	27.8	27.8	1.74
B	2	10,000	22.2	50.0	3.12
C	4	25,000	111.1	161.1	10.07
D	3	40,000	133.3	294.4	18.40

The Gantt Chart in Figure 3-2 shows the same information as the above data.

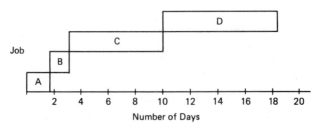

FIGURE 3-2 A picture of a machine's or department's time schedule of work.

This plant operates a single machine 16 hours per day, five days per week. There are 294.4 hours of backlog at 16 hours per day = 18.4 days of work in the backlog. What if a customer comes in with a job and wants it in ten days? The job is estimated to take only 48 hours of machine time. Can you deliver? What about the other four jobs? When have you promised them?

One scheduling philosophy is that operating departments are compared to buckets of time. The size of the bucket is the number of hours that each department can produce in a 24-hour day. Example:

Dept.	No. of Machines	Two Shifts Hours per Day Available	Historical Department Performance	Hours Capacity
Shears	2	32	85	27.2
Presses	6	96	90	86.4
Press breaks	4	64	80	51.2
Welding	4	64	75	48.0
Paint	3	48	95	45.6
Assembly line	1	80	90	72.0

Therefore, the scheduler can keep adding work to any department for a specific day until the hour capacity is reached; then it spills over to the next day.

Without good time standards, manufacturing management would have to carry great quantities of inventory not to run out of parts. Inventory is a great cost in manufacturing; therefore, knowledge of time standards will reduce inventory requirements, which will reduce cost. Production inventory control is an area of major importance in industrial technology programs, and time study should be a prerequisite.

As we've said before, how could you schedule a plant without time standards? Anything else is a guess.

5. How do we balance the work load among people on assembly and packout lines?

The objective of assembly line balancing is to give each operator as close to the same amount of work as possible. This can only be accomplished by breaking down the tasks into the basic motions required to do every single piece of work and reassembling the work into jobs of near equal time value. The work station or stations with the largest time requirement is designated as the 100% station and limits the output of the assembly line. If an industrial technologist wants to improve the assembly line (reduce costs), he or she would concentrate on the 100% station. If we reduce the 100% station in the following example by 1%, we save 4% of overall assembly line labor cost.

Example: Assembly Line Balance

Operator No.	Operation Description	Standard Time in Minutes	Load %	Pieces per Hour	Hours per Piece
1	Asemble 1,2,3	.250	84	200	.005
2	Assemble 4,5,6	.300[a]	100[a]	200	.005
3	Assemble 7,8,9	.275	92	200	.005
4	Assemble 10,11,12	.225	75	200	.005
				Total hours	.020[b]

[a]The busiest station on this line is work station 2, with .300 minutes of work. As soon as we identify the busiest station, we identify it as the 100% station, and by highlighting this standard and the 100% load, we communicate that only this standard is important. Even though they could work faster, every work station is limited to 200 pieces per hour, because station 2 is limiting the output of the whole assembly line.

[b].020 is the total hours required to assemble one finished unit. If we multiply this total hours by the average assembly wage rate, we have the total assembly labor cost. A better line balance is a lower total hours.

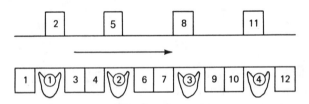

FIGURE 3-3 Assembly line layout based on assembly line balance.

See Figure 3-3 for an example of assembly line layout based on assembly line balance.

How else could we divide the work load equally and fairly without time standards? Any other way would be a guess.

Line balancing is the subject of chapter 15.

6. How do we measure productivity?

Productivity is a measure of output divided by input. If we are talking about labor productivity, then we are developing a number of units of production per manhour.

Example:

$$\text{Present} = \frac{\text{Output} = 1{,}000 \text{ units/day}}{\text{Input} = 50 \text{ people @ 8 hours/day}} = \frac{1{,}000}{400} = 2.5 \text{ units per manhour}$$

$$\text{Improved} = \frac{\text{Output} = 2{,}000 \text{ units/day}}{\text{Input} = 50 \text{ people @ 8 hours/day}} = \frac{2{,}000}{400} = 5.0 \text{ units per manhour}$$

or a 100% increase in productivity (a doubling of production).

We could also increase productivity by maintaining the output constant or reduce the number of people.

$$\text{Improved output} = \frac{1,000 \text{ units/day}}{40 \text{ people @ 8 hours/day}} = \frac{1,000}{320} = 3.2 \text{ units per manhour.}$$

These examples are good for plants or whole industries, but for individuals we use

$$\frac{\text{Earned hours}}{\text{Actual hours}} = \% \text{ performance.}$$

Earned hours are the hours of work earned by the operator based on the work standard and the number of pieces produced by the operator. For example, if an operator worked eight hours and produced 1,000 units on a job with a time standard of 100 pieces per hour, we would have the following:

A. Earned hours $= \dfrac{1,000 \text{ pieces produced}}{100 \text{ pieces per hour}} = 10 \text{ hours}$

B. Actual hours $= 8$ hours

Actual hours are the real time the operator works on the job (also called the time clock hours).

C. % performance $= \dfrac{\text{Earned hours} = 10}{\text{Actual hours} = 8} = 125\%.$

Industrial technologists will improve productivity by reporting performances of every operation, operator, supervisor, and production manager every day, week, month, and even yearly.

Performance reports are based on daily time cards filled out by operators and extended by time keepers.

Performance control systems improve productivity by 42% on the average. Companies without performance control systems typically work at 60% performance. Once time standards are set and a daily performance report is installed, production will increase to an average of 85% performance. This is accomplished by

1. Identifying nonproductive time and eliminating it
2. Identifying poorly maintained equipment and fixing it
3. Identifying causes for downtime and eliminating them
4. Planning ahead for the next job

We hold problems up to the light of day and fix the problems. In plants that do not have standards, the employees know no one cares how much they produce. Management's reaction to problems speaks louder than their words. How can supervisors know who is producing and who is not if they don't have standards? How would management know the magnitude of problems such as downtime for lack of maintenance, material, instruction, tooling, services, etc., if downtime is not reported?

We discuss the performance control system in chapter 12.

7. How can we pay our people for outstanding performance?

Every manufacturing manager would like to be able to reward outstanding employees. Every supervisor knows who he or she can count on to get the job done. Yet only about 25% of production employees have an opportunity to earn increased pay for increased output. A 1980, 400-plant study by an industrial engineering consultant, Mitchell Fein, found that when employees are paid via incentive systems, their performance improves by 41% over measured work plans, and 65% over no standards or no performance control system.

Stage I   Plants with no standards operate at 60% performance.

Stage II   Plans with standards and a performance control system operate at 85% performance.

Stage III Plants with incentive systems operate at 120% performance.

A small company with 100 employees could save about $750,000 per year on labor costs going from no standards to incentive systems, and the employees will take home 20% more pay.

A National Science Foundation study found that when workers' pay was tied to their efforts,

1. Productivity improved;
2. Cost was reduced;
3. Workers' pay increased; and
4. Workers' morale improved.

Everyone can win with incentive systems. What is management waiting for? Chapter 13 discusses group and individual incentive systems.

8. How can we select the best method or evaluate cost reduction ideas?

A basic rule of production management is that all expenditures must be cost justified. Second, a basic rule of life is that everything changes. We must keep improving or become obsolete. To justify all expenditures, the savings must be calculated. This is called the *return*. The cost of making the change is also counted. This is called the *investment*. When the return is divided by the investment, the resulting ratio indicates the desirability of the project. This ratio is called the ROI, or return on investment. To provide a uniform method of evaluating ROI, annual savings are used; therefore, all percentages are per year.

Example: We have been producing product A for several years and look forward to several more years of sales at 500,000 units per year or 2,000 units per day. The present method requires a standard time of 2.0 minutes per unit or 30 pieces per hour. At this rate, it takes 33.33 hours to make 1,000 units. All production will run on the day shift.

A. Present Method and Costs With a labor rate of $10.00/hour, the labor cost will be $333.30 to produce 1,000 units. The cost of 500,000 units per year would be $166,665.00 in direct labor.

$$\frac{1,000 \text{ pieces}}{30 \text{ pieces per hour}} = 33.33 \text{ hours}/1,000 \text{ units}$$

B. New Method and Costs We have a cost reduction idea. If we buy this new machine attachment for $1,000, the new time standard would be lowered to 1.5 minutes per unit. Will this investment be good for us?

First, how many attachments will we have to buy to produce 500,000 units per year?

$$\frac{500,000 \text{ units/year}}{250 \text{ days/year}} = 2,000 \text{ units/day}$$

$$\begin{array}{rl}
480 & \text{minutes/shift} \\
-50 & \text{minutes/shift downtime} \\
\hline
430 & \text{minutes/shift @ 100\%} \\
@80\% & \text{expected efficiency} \\
\hline
344 & \text{effective minutes available to produce 2,000 units/shift}
\end{array}$$

$$\frac{344 \text{ minutes}}{2,000 \text{ units}} = .172 \text{ minutes/unit}$$

To produce 2,000 units per shift, we need a part every .172 minutes.

$$\text{Number of machines} = \frac{1.50 \text{ minutes/cycle}}{.172 \text{ minutes/unit}} = 8.7 \text{ machines}$$

We will purchase 9 attachments at $1,000 each. Our investment will be $9,000 (9 × 1,000). Second, what is our labor cost?

$$\text{Pieces per hour} = \frac{60 \text{ minutes/hour}}{1.5 \text{ minutes/part}} = 40 \text{ parts/hour or 25 hours}/1,000$$

25 hours/1,000 × $10.00/hour wage rate = $250/1,000

500,000 units will cost 500 × $250 = $125,000

New labor costs will be $125,000 per year.

C. Savings: Direct Labor Dollars

Old method $166,665 per year

New method $125,000 per year

Savings $41,665 per year

$$\frac{\text{Return (savings) } \$41,665/\text{year}}{\text{Investment (cost) } 9,000} = 463\%$$

$$\text{ROI} = 463\%$$

D. Return on Investment This investment will pay for itself in less than three months. If you were the manager, would you approve this investment? Of course you would, and so would anyone.

Cost reduction programs are important to the well-being of a company and the peace of mind of the industrial engineering department. A department that shows a savings of $100,000 per employee per year doesn't have to worry about layoffs or elimination. A properly documented cost reduction program will update all time standards as soon as methods are changed. Every standard affected must be changed immediately.

Cost reduction calculations can be a little more complicated than our example, which did not account for

1. Taxes
2. Depreciation
3. Time value of money
4. Surplus machinery—trade in
5. Scrap value.

9. How do you evaluate new equipment purchases to justify their expense?

The answer to this question is the same as the answer to Question 8. Every new machine is a cost reduction. No other reason is acceptable.

10. How do we develop a manpower budget?

This question was answered in Question 2 when we determined the number of people to hire. Budgeting is one of the most important management tools, and one must understand it fully to become an effective manager. It is said that you become a manager when you are responsible for a budget. You are a promotable manager when you come in under budget at the end of the year. Budgeting is a part of the cost-estimating process. Labor is only one part of the budget, but it is one of the most difficult to estimate and control. How would we do it without time standards? It would be a very expensive guess.

How can managers guess at such important decisions as asked in this chapter? Much of manufacturing management has received no formal training in making these decisions. It will be your job to show them the scientific way to manage their operations.

QUESTIONS

1. What are time standards used for?
2. What is the definition of a time standard?
3. What three numbers make up a time standard?
4. What is productivity? How is it measured for individuals?
5. What are the three basic levels of productivity?
6. How many machines should you buy and how many people should you hire if 3,000 units are needed per shift in a 75% efficient plant that has 10% downtime? The machine time standard is .284 minutes. How much will a unit cost to produce if the operator earns $7.50 per hour? How many units will be produced per shift?

CHAPTER 4

Techniques of Motion and Time Study

Technology is the application of techniques, as discussed in chapter 1. The techniques of motion and time study are tools for the technologist to use to improve the operations of plants. The more techniques a technologist has at his or her command, the more valuable that technologist will be to the company.

Motion and time study covers a wide variety of situations. At one time a job must be designed, work stations and machines built, and a time standard set before the plant is built. In this situation, PTSS (predetermined time standards system) or standard data would be the techniques used to set the time standard. Once a machine or work station has been operated for a while, the stopwatch technique is used. Some jobs occur once or twice a week, while others repeat thousands of times per day. Some jobs are very fast, while others take hours. Which technique do we use? The job of an industrial technologist is to choose the correct technique for each situation and correctly apply that technique.

TIME STANDARDS TECHNIQUES

Five techniques of time standard development are studied in this text:

1. Predetermined time standard systems
2. Stopwatch time study
3. Work sampling
4. Standard data
5. Expert opinion standard and historic data.

A brief description of these five techniques is presented in this chapter to give the student direction. Each technique will be developed fully in its own chapter.

Predetermined Time Standards Systems (PTSS)

When a time standard is needed during the planning phase of a new product development program, the PTSS technique is used. At this stage of new product development, only sketchy information is available, and the technologist must visualize what is needed in the way of tools, equipment, and work methods. The technologist would design a work station for each step of the new product manufacturing plan. Each work station would be designed, a motion pattern would be developed, each motion would be measured, a time value assigned, and the total of these time values would be the standard. This time standard would be used to determine the equipment, space, and people needs of the new product and its selling price.

Frank and Lillian Gilbreth developed the basic philosophy of predetermined time motion systems. They divided work into seventeen work elements:

1. Transport empty
2. Search
3. Select
4. Grasp
5. Transport loaded
6. Preposition
7. Position
8. Assemble
9. Disassemble
10. Release load
11. Use
12. Hold
13. Inspect
14. Avoidable delay
15. Unavoidable delay
16. Plan
17. Rest to overcome fatigue.

These seventeen work elements, as mentioned in chapter 2, are known as *therbligs*. Each therblig was reduced to a time table, and when totaled, a time standard for that set of motions was determined.

Methods time measurement (MTM)* and work factors are two popular predetermined time systems inspired by the Gilbreths' work.

MTM, developed by Maynard, Stegemarten, and Schwab in 1948, is probably the best known predetermined time system in use today. MTM-1* has ten elements of micromotion. Each element is assigned a number of time-measured units (TMU), which are in .00001 hours (one hundred thousandths of an hour). One hour equals 100,000 TMUs. One minute equals 1,667 TMUs. The elements of MTM are very similar to PTSS, but PTSS is in .001 minutes, or one thousandths of a minute.

MTM-2 and MTM-3 are faster but less accurate systems of MTM. MTM-1 takes 350 times the cycle time to analyze the job, whereas MTM-2 takes 150 times the cycle time and MTM-3 take 50 times (Magnusson, 1972). MTM-1 accuracy is within ±7%, whereas MTM-3 is about ±20% accurate.

Newer computer-based MTM data are available. H. B. Maynard and Company Inc. has an advanced technique called MOST (Maynard Operational Sequence Technique). MOST is four times faster than MTM-3 (Kjell B. Zandin, 1980).

The MTM association and qualified individuals worldwide conduct training programs of up to two weeks. The MTM blue card is awarded to graduates of these programs and is recognition of a professional level of competency in job design and time standards development. MTM training will help your career, but the time required to take their course, the registration fees, and the temporary living expenses may require the average student to wait until employed by a company using the system.

PTSS was developed from MTM and other predetermined time systems for the expressed reason to teach a system within a few hours. PTSS is a simplified system. It is a good system, but additional training will be desirable. If you understand the concepts of PTSS, you will have a head start in learning any other system.

Figure 4-1 shows an example of a PTSS. PTSS is the first technique of time study covered in this text, and it is the last technique in the methods chapters (chapter 8) because it is both a methods and a time study technique.

Stopwatch Time Study

Stopwatch time study is the method most manufacturing employees think of when talking about time standards. Fredrich W. Taylor started using the stopwatch around 1880 for studying work. Because of its long history, this technique is a part of many union contracts with manufacturing companies.

Time study is defined as the process of determining the time required by a skilled, well-trained operator working at a normal pace doing a specific task. Several types of stopwatches could be used:

1. Snapback: in one hundredths of a minute

2. Continuous: in one hundredths of a minute

*MTM refers to all MTM systems—MTM 1, MTM 2, MTM 3 and MTM 4.

FRED MEYERS & ASSOCIATES — PREDETERMINED TIME STANDARDS ANALYSIS

OPERATION NO. 25   PART NO. 2220   OPERATION DESCRIPTION: Assemble and bolt 2 flanges to body.
DATE: 3-5-xx   TIME:
BY I.E. Meyers

DESCRIPTION-LEFT HAND	FREQ	LH	TIME	RH	FREQ	DESCRIPTION-RIGHT HAND	ELEMENT TIME
To next body		R30	18	M30		Aside completed	
Grasp		G2	6	RL		Release	
Move to fixture		M30	18	R30		To L.H.	
			2	G2		Grasp body in L.H.	
In fixture		AP1	5	AP1		Into fixture	
			49				.049
Get & Assemble Brackets							
		R12	9	R12		To R.H. Bracket	
		G2	6	R2		To R.H. Bracket	
			4 / 6	G2		Grasp Bracket	
Same as R.H.		M12	9	M12		To body	
		AP2	10				
			5	AP2		On body	
		RL	--	RL			
			49				.049
Get & Assemble 4 bolts - Hand tight							
		R8	7	R8		To Bolts	
		G3	9				
Same as L.H.			9	G3		Grasp Bolts	
		M8	7	M8		To Fixture	
		AP1	5	Ap1		In Body	
	10	G4	40	G4	10	Turn 10 Times	
		SF	5	SF		Tighten	
			81		2	3rd & 4th Bolt	.162

TIME STUDY CYCLE		COST:			TOTAL NORMAL TIME IN MINUTES PER UNIT	.260
.25	1.20	2.11				
.50	1.42	2.35	HOURS PER UNIT	.00477	+ 10 % ALLOWANCE	.026
.74	1.65	2				
.97	1.89		DOLLARS PER HOUR	8.50	STANDARD TIME	.286
TOTAL		2.35 / 10				
OCC			DOLLARS PER UNIT	$ .041	HOURS PER UNIT	0 0 4 7 7
AVG. OCC		.235				
LEV FACT		110			PIECES PER HOUR	209
NORM. TIME		.258				

FIGURE 4-1 PTSS example.

PTSS side B example.

3. Three watch: continuous watches

4. Digital: in one thousandths of a minute

5. TMU (time-measured unit) in one hundred thousandths of an hour

6. Computer: in one thousandths of a minute.

All but the TMU watch read in decimal minutes. The TMU watch reads in decimal hours. Digital watches and the computer are much more accurate, but tradition favors the first three watches.

Nearly 25% of this motion and time study book is dedicated to the stopwatch time study technique because of its traditional importance.

Two different time study procedures are covered in this book:

1. Continuous time study (Figure 4-2 is used on short duration jobs)

2. Long cycle time study (see Figure 4-3).

Long cycle time study may be used for either very long jobs (30 minutes or more) or 8-hour studies, or for jobs where the elements are often performed out of sequence.

The 8-hour time study is used to find out what causes an operation's poor perfor-

FRED MEYERS & ASSOCIATES — TIME STUDY WORKSHEET

[ ] SNAP BACK [X] CONTINUOUS

OPERATION DESCRIPTION: ASSEMBLE PARTS 2 & 4, MACHINE SCREW & STAKE. INSPECT

PART NUMBER 4650-0950	OPERATION NO. 1515	DRAWING NO. 4650-0950	MACHINE NAME PRESS	MACHINE NUMBER 21	[X] QUALITY OK ?
OPERATOR NAME MEYERS	MONTHS ON JOB 5	DEPARTMENT ASSEMBLY	TOOL NUMBER M61	FEEDS & SPEEDS. NONE	[X] SAFETY CHECKED ?
PART DESCRIPTION: GOLF CLUB SOLE ASSEMBLY - WOOD & STEEL		MATERIAL SPECIFICATIONS:		MACHINE CYCLE .030 TIME 8:30 AM.	[X] SETUP PROPER ? NOTES:

READINGS / time study data:

ELEMENT #	ELEMENT DESCRIPTION	R/E	1	2	3	4	5	6	7	8	9	10	TOTAL/CYCLES	AVERAGE TIME	% R	NORMAL TIME	FREQUENCY	UNIT NORMAL TIME	RANGE	R/X̄	HIGHEST
1	ASSEMBLY	R	9	41	71	1.07	38	77	2.08	48	77	3.07	.76	.084	90	.076	1/1	.076	.02		✓
		E	.09	.09	.09	(15)	.08	.08	.10	.07	.08	.08	/9								
2	DRIVE SCREW	R	15	46	79	13	43	82	14	53	82	93	.51	.057	100	.057	1/1	.057	.03	.53	✓
		E	06	05	08	06	05	05	06	05	05	(10)	/9								
3	PRESS	R	28	59	94	27	66	95	28	66	96	4.06	1.22	.136	110	.150	1/1	.150	.02		
		E	13	13	15	14	(23)	13	14	13	14	13	/9								
4	INSPECT	R	32	62	92	30	69	98	41	69	99	4.09	.25	.031	100	.031	1/1	.031	.01		
		E	.04	.03	(-.02)	.03	.03	.03	(13)	.03	.03	.03	/8								
5	LOAD SCREWS	R										3.83	.76	.76	125	.950	1/10	.095	—		
		E					*1	*2			*3	.76	1								

FOREIGN ELEMENTS:
*1 .23 PART JAMMED.
*2 .13 TRIED TO REWORK PART.
*3 .10 RESTART FROM LOADING SCREWS.

ENGINEER: FRED MEYERS   DATE: 10 / 10 / XX
APPROVED BY: FRED MEYERS   DATE: 10 / 10 / XX

NOTES:
LOAD SCREWS COULD BE IMPROVED
TO ELIMINATE .095 MIN.

```
.409
-.095      (SAVE)
.314       .00750
+.031      -.0575
.345       .00275 Hrs/Unit
           X $10.00/Hr.
.00575 Hrs .0575 $/Unit
174 Pieces/Hrs  500,000/Yr.
                #28,750
```

R/X̄	# CYCLES
.1	2
.2	7
.3	15
.4	27
.5	42
.6	48 61
.7	83
.8	108
.9	138
1.0	169

TOTAL NORMAL MIN. .409
ALLOWANCE + _____ 10 % .041
STANDARD MINUTES .450
HOURS PER UNIT .00750
UNITS PER HOUR 133

ON BACK
WORK STATION LAYOUT
PRODUCT SKETCH

FIGURE 4-2 Time study example: Continuous form.

mance. Figure 4-4 shows an analysis of what caused a quart canning line to be down. Each line on the analysis represents 1 hour. Each line is divided into sixty parts (1 minute each). An X through five minutes means the machine was down for five minutes. The number above the X indicates the reason for the downtime. At the end of the study, the analysis shows how many times the machine stopped for each reason and the total downtime for that reason. This total downtime can be converted easily to total dollars, and we have an indication of how much money we can spend to solve each problem.

Work Sampling

Work sampling is the same scientific process used in Neilson ratings, Gallup polls, attitude surveys, and federal unemployment statistics. We observe people working and draw conclusions. Everyone who has ever worked with someone else has done work sampling; you have an opinion of how hard this other person works:

FRED MEYERS & ASSOCIATES LONG CYCLE TIME STUDY WORK SHEET

PART NO. Quart Line | OPERATION DESCRIPTION: 4 people (loader, Machine,
OPERATION NO. Line | Cartons, Unloader) Automatic Quart Line Cann
DATE/TIME 10/10/xx | MACHINE: TOOLS, JIGS: #1--300 CANS/Minute
BY I.E. Meyers | MATERIAL: Motor oil any weight

ELEMENT #	ELEMENT DESCRIPTION Started 7:00 AM Ended 3:30 PM	ENDING WATCH READING	ELEMENT TIME	% R	NORM. TIME
1	Shift Start up--No production	7:05	5.0	100	5.0
2	Run	7:06	1.00	100	1.0
3	Stopped--operator forgot something	7:07½	1½	0	----
4	Run	7:14	6½	100	6.5
5	No Lids	7:16½	2½	0	----
6	Run	7:19	2½	100	2.5
7	Check Temperature	7:20½	1½	70	1.05
8	Run	24	3½	100	3.5
9	Box Jammed in Former	25	1	110	1.1
10	Run	28¼	3¼	100	3.25
11	Bad Can	29	3/4	120	.9
12	Palletizer Jam-Bad Pallet	31	2	110	2.2
13	Run	33	2	100	2
14	Bad Box in Former	7:33½	½	130	.65
15	Run	7:41	7½	100	7.5
16	Bad Box in Former	7:41½	½	140	.7
17	Run	7:51	9½	100	9.5
18	Bad Box in Former	7:54	3	120	3.6
19	Run	7:56	2	100	2.0
20	No Lids in Machine	8:00½	4½	0	----

FIGURE 4-3 Time study example: Long cycle time study worksheet, page 1 of 8.

1. "Every time I look at him, he's working"; or

2. "He's never working"; or

3. Somewhere in between.

Supervisors, using informal work sampling, are forming attitudes of employees all the time.

Industrial technologists can walk through a plant and state, "This plant is working at 75% performance." They should continue to say ± 10% or so, depending on how many people they observed (number of samples). You could walk through a plant of 250 people one time and count people who are working and those that are not working and calculate the performance of that plant within ± 10%. Industrial engineering con-

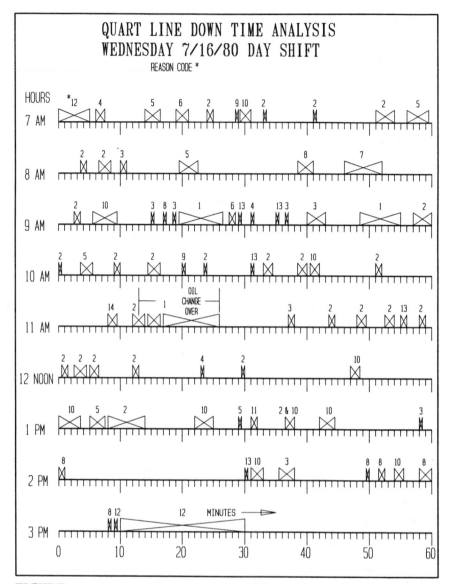

FIGURE 4-4 Graphical analysis of 8-hour time study.

sultants often start their consulting proposal with such statistics. Consultants expect to find 60% performance in plants without standards, but that is an average. A specific plant may have better management and be averaging between 70 and 75%. They could not save as much in this plant.

    Setting standards using work sampling is not very difficult. The industrial technologist samples a department and finds the following statistics:

Task	No. of Observations	% Total	Hours Worked	Pieces Produced	Pieces/Hour[a]
Assemble	2,500	62.5%	625	5,000[b]	8
Idle	1,500	37.5%	375	—	
Total	4,000	100.0%	1,000[c]		

[a] Pieces per hour $= \dfrac{\text{pieces} = 5,000}{\text{hours} = 625} = 8$ pieces per hour.

[b] From supervisor (number finished products put in the warehouse).

[c] From payroll (hours paid during our study).

Eight pieces per hour is not quite the time standard. We haven't added allowances. How much time is in the 625 hours for breaks, scheduled or unscheduled? How much time is in there for delays? None. Actual hours worked is 625. All other nonwork time is part of the 375 hours which we throw away. We could add an appropriate amount of extra time to cover personal time, fatigue time, or delay time. This extra time is called *allowances*. Ten percent extra time is considered normal. A time standard of 7.2 pieces per hour would be appropriate.

Chapter 11 includes a step-by-step procedure for conducting a full work sampling study.

Standard Data

Standard data should be the objective of every motion and time study department. Standard data is the fastest and cheapest technique of setting time standards, and standard data can be more accurate and consistent. Starting with many previously set time standards, the industrial technologist tries to figure out what causes the time to vary from one job to another on a specific machine or class of machine. For example, walking time would be directly proportional to the number of feet, paces, yards, or meters walked. There might be two curves on the graph: obstructed and unobstructed.

A second example is counting playing cards. Time for counting cards would be directly proportional to the number of cards counted. Can you think of any other reasons for the time to vary?

There are several ways of communicating the time standard to future generations of factory workers, supervisors, and engineers:

1. Graph (see Figure 4-5)
2. Table
3. Worksheet (again, see Figure 4-5)
4. Formula.

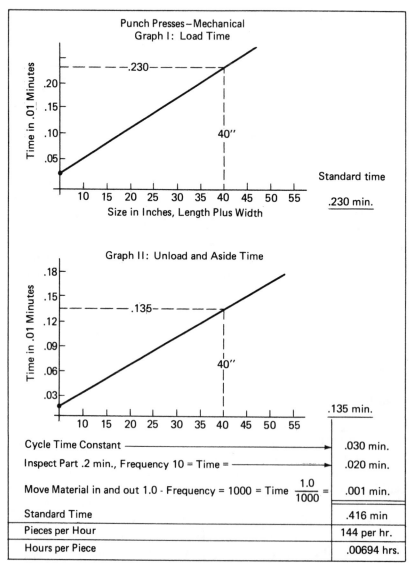

FIGURE 4-5 Standard data worksheet example.

Metal-cutting machines are examples of the need for and use of formulas. A blueprint calls for the drilling of a hole through the steel plate. Three pieces of information are required:

1. What size hole?
2. What is the material?
3. What tool do we use?

With this information, we look up the feeds and speeds in the *Machinery's Handbook*.*
Feeds and speeds are communicated as follows:

Speed   500 feet/min.

Feed   .002 inches per revolution.

By substituting this information into three simple formulas, we determine the time standard.

Other machines, like welders, have simpler formulas, such as 12″ per minute. The machine manufacturers are a good source of standard data.

Expert Opinion Time Standards and Historic Data

An expert opinion time standard is an estimation of the time required to do a specific job. This estimate is made by a person with a great experience base. Many people say, "You can't set time standards on my work." A good industrial technologist's response would be, "You are right, but I know someone who can—you can!" The one-of-a-kind nature of many staff and service workers makes setting time standards with the more traditional techniques unprofitable. Engineering maintenance and some office workers never do the same thing twice, but goals are still needed (time standards). Maintenance work is controlled by work order. Why not ask an expert how long this requested work will take? In well-managed companies, new maintenance projects will not be approved until the job is estimated. These time standards would be used to schedule and control maintenance work just as you would schedule and control the work performed by a machine operator.

The expert in an expert opinion time standards system is usually a supervisor. In larger departments, a specialist may be used. For example, in the maintenance department, the position would be called a maintenance planner. The expert would estimate every job and maintain a backlog of work. The backlog of work would give the department time to plan the job, thereby performing that job more effectively.

Expert opinion time standards and the backlog control system are discussed in detail in chapter 12.

Historic data is an accounting approach to expert opinion time standard systems. A record is kept of how much time was used on each job. When a new job comes along, it is compared to a previous job standard. These standards are then used in a labor performance control system. The problem with historic time standards is that they do not reflect the time the job should have taken. Inefficiency is built into such a system, but a bad standard is better than no standard at all.

Figure 4-6 will help you choose the correct technique for setting time standards

---

Machinery's Handbook, The Industrial Press, New York, New York.

FIGURE 4-6 Which time standards technique do we use?

Cycle Time	High 1000's	Medium 100's	Low 10's
		Volume of Production	
Long	Work sampling	Work sampling Stopwatch	Expert opinion Work sampling History
Medium	Work sampling Stopwatch PTSS	Stopwatch Work sampling	Expert opinion History Stopwatch
Short	PTSS	PTSS Stopwatch	Stopwatch Expert opinion

Note: Standard data is the ultimate time standard technique and can be used in all situations. Standard data should be the goal of all time study departments.

FIGURE 4-7 Flow diagram: The path taken by a part as it flows through a plant.

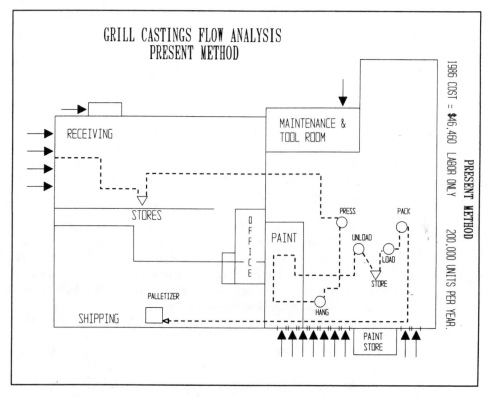

FRED MEYERS & ASSOCIATES — PROCESS CHART

☒ PRESENT METHOD ☐ PROPOSED METHOD   DATE: 5/6   PAGE 1 OF ___

PART DESCRIPTION: 2,000 UNITS/SHIFT
GRILL CASTING 75102

OPERATION DESCRIPTION: FROM RECEIVING TO SHIPPING

SUMMARY	PRESENT NO.	PRESENT TIME	PROPOSED NO.	PROPOSED TIME	DIFF. NO.	DIFF. TIME	ANALYSIS:	FLOW DIAGRAM
○ OPERATIONS							WHY / WHEN	ATTACHED
⇨ TRANSPORT.							WHAT / WHO	(IMPORTANT)
☐ INSPECTIONS							WHERE / HOW	
D DELAYS							STUDIED BY: F. MEYERS	
▽ STORAGES								$7.00 PER HR.
DIST. TRAVELED	FT.		FT.					

STEP	DETAILS OF PROCESS	METHOD	OPERATION	TRANSPORT	INSPECTION	DELAY	STORAGE	DISTANCE IN FEET	QUANTITY	TIME (.00001)	COST PER UNIT	TIME/COST CALCULATIONS
1	RECEIVING UNLOAD TRUCK	PALLET FORK	●						120	31	.0025	2 MIN/PALLET
2	MOVE TO STORES	FORK TRUCK		➡				125'	120	23	.0016	2.5 MIN/PALLET
3	STORAGE	RACK					▼	40,000				30 DAYS $3.00 EACH
4	MOVE TO MACHINE	FORK TRUCK		➡				625'	120	55	.0039	12.5 MIN/PALLET
5	WAIT AT MACHINE					D						30 MINUTES
6	PUNCH		●							532	.0372	188 PER HR.
7	WAIT					D				62	.0043	30 MINUTES
8	MOVE TO PAINT	FORK TRUCK		➡				200'	120	3	.0002	4.0 MINUTES
9	WAIT					D			INV. ⬦			
10	PLACE ON CONVEYOR	HAND	●							595	.0417	336 PER HR.
11	TO PAINT	CONVEYOR		➡				10'			FREE	
12	HANG ON LINE		●							298	.0209	336 PER HR.
13	CLEAN-PAINT-BAKE	CONVEYOR	●	➡				400'		(INDIRECT) FREE		
14	UNLOAD		●							298	.0209	336 PER HR.
15	STACK		●							298	.0209	336 PER HR.
16	MOVE TO STORAGE	HAND		➡				20'		298	.0209	336 PER HR.
17	STORE FOR LINE						▼		INV. ⬦			

FIGURE 4-8 Process chart: Record of everything that happens to a part as it flows through a plant.

MOTION STUDY TECHNIQUES

The following are motion study techniques:

1. Flow diagram (see Figure 4-7)
2. Operations chart (see Figure 3-1)
3. Process chart (see Figures 4-8 and 4-9)
4. Flow process chart
5. Operations analysis chart

STEP	DETAILS OF (PRESENT/PROPOSED) METHOD	METHOD	OPERATION	TRANSPORT	INSPECTION	DELAY	STORAGE	DISTANCE IN FEET	QUANTITY	TIME .0001	COST PER UNIT	TIME/COST CALCULATIONS
18	TO CONVEYOR	HAND	○	➡	☐	D	▽	20'		400	.0280	260 PER HR.
19	TO PACKOUT LINE	CONVEYOR	○	➡	☐	D	▽	20'				
20	PACK IN CARTON		●	⇨	☐	D	▽			400	.0280	260 PER HR.
21	TO STORES	CONVEYOR	○	➡	☐	D	▽				FREE	
22			○	⇨	☐	D	▽					
23			○	⇨	☐	D	▽					
24			○	⇨	☐	D	▽					
25			○	⇨	☐	D	▽					
26			○	⇨	☐	D	▽					
27			○	⇨	☐	D	▽					
28			○	⇨	☐	D	▽					
29			○	⇨	☐	D	▽					
30			○	⇨	☐	D	▽					
31			○	⇨	☐	D	▽					
32			○	⇨	☐	D	▽					
33			○	⇨	☐	D	▽					
34			○	⇨	☐	D	▽					
35			○	⇨	☐	D	▽					
36			○	⇨	☐	D	▽					
37			○	⇨	☐	D	▽					
38			○	⇨	☐	D	▽					
39			○	⇨	☐	D	▽					
40			○	⇨	☐	D	▽					
41			○	⇨	☐	D	▽					
42			○	⇨	☐	D	▽					

FIGURE 4-9 Process chart: page 2.

6. Operator/machine chart
7. Gang chart
8. Multimachine chart
9. Motion patterns
10. Predetermined time standards system.

These techniques are divided into two groups. Group one covered in chapter 5 are the techniques for studying the entire plant and product. These techniques are called flow analysis tools. Techniques 1 to 4 are these techniques.

Group two is discussed in chapters 6, 7, and 8. These techniques are 5 to 10 above. These techniques are individual operations design.

The predetermined time standards system (PTSS) is an outstanding methods analysis tool for an individual work station and will allow for the development of the best method. A by-product of PTSS is a time standard, so PTSS is the first time standards technique we discuss in this text.

Motion study is learned before time study because time standards should not be set for poor methods. Great savings can be generated from the application of motion study techniques. The principles of motion economy will help us find the best method.

Learn the techniques of motion and time study, and you will always be in demand.

QUESTIONS

1. What are the five techniques of setting time standards?
2. Which technique would be used when no work station is available?
3. Which technique is the most popular?
4. Which technique would be used for maintenance work?
5. Which is the best technique for setting time standards?
6. Which technique is both a methods and time study technique?

CHAPTER 5

Techniques of Methods Design: The Broad View

INTRODUCTION

Prior to studying individual jobs, the technologist should study the overall flow of a product through the facility. Understanding as much as possible about the present condition prepares us to improve that condition. In the case of a product to be manufactured, we take that product apart and study the manufacturing sequence of each part and the sequence of assembly of the parts into subassembly, finished product, and packed out. Techniques, discussed in this chapter, are required to show and tell all the information required to build a manufacturing facility complete with the proper number of people, machines, and tools. These techniques will lead the technologist to ask the right questions to improve even the newly conceived plan. These techniques force consideration of the best method of doing a job.

The techniques covered in this chapter will be used for overall product flow study. Only one reason justifies the effort of motion study—cost reduction. When a technician finishes a technique, he or she turns around and does it again—only better. As a new

COST REDUCTION FORMULA

Ask These Questions	For Each	To Seek These Results
Why	Operation	Eliminate
What	Transportation	Combine
When	Storage	
Who		
Where	Inspection	Reroute
How	Delay	Simplify

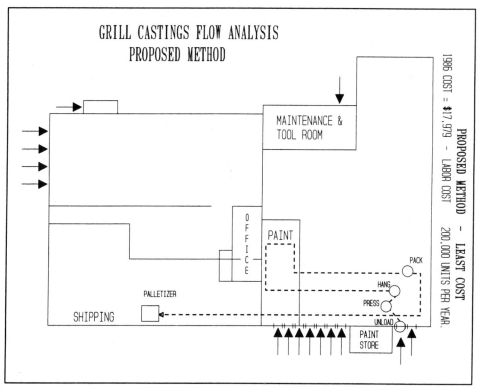

FIGURE 5-1 Flow diagram: An improvement over Figure 4-7.

technologist, once you have applied these techniques, you should know as much about your company's manufacturing systems as anyone.

FLOW DIAGRAMS

The flow diagram (see Figure 5-1) shows the path traveled by each part from receiving to stores to fabrication of each part to subassembly to final assembly to packout to warehousing to shipping. These paths are drawn on a layout of the plant.

The flow diagram will point out problems with such things as the following:

1. Cross Traffic

Cross traffic is where flow lines cross. Cross traffic is undesirable, and a better layout would have less intersecting paths. Anywhere traffic crosses is a problem because of congestion and safety considerations. Proper placement of equipment, service, and departments will eliminate most cross traffic.

2. Backtracking

Backtracking is material moving backward in the plant. Material should always move toward the shipping end of the plant. If it is moving toward receiving, it's moving backward. Backtracking costs three times as much as flowing correctly. For example, five departments have flow like this:

How many times did material move between departments 3 and 4? Three times: twice forward and once backward. If we rearranged this plant and changed departments 3 and 4 around, we would have straight through flow with no backtracking:

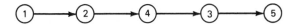

We travel less distance. In the former example, we traveled six blocks (a block is one step between departments next to each other). In the straight line flow, we traveled only four blocks—a 33% increase in productivity.

3. Distance Traveled

It costs money to travel distance. The less distance traveled, the better. The flow diagram is developed on a layout, and the layout can be easily scaled and distance of travel calculated. By rearranging machines or departments, we may be able to reduce the distances traveled.

Because flow diagrams are created on plant layouts, no standard form is used. There are few conventions to restrict the designer. The objective is to show all the distances traveled by each part and to find ways of reducing the overall distance.

The flow diagram is developed from routing sheet information, assembly line balance, and blueprints. The routing sheet specifies the fabrication sequence for each part of a product. The sequence of steps required to make a part is practical and has some room for flexibility. One step may come before or after another step, depending on conditions. The sequence of steps should be changed to meet the layout, if possible, because that requires only a paperwork change. But if the sequence of operations cannot be changed and the flow diagram shows backtracking, moving equipment may be necessary. Our objective will always be to make a quality part the cheapest, most efficient way possible.

STEP-BY-STEP PROCEDURE FOR DEVELOPING A FLOW DIAGRAM

Step 1: The flow diagram starts with an existing or proposed scaled layout.

Step 2: From the route sheet, each step in the fabrication of each part is plotted and connected with a line, color code, or other method of distinguishing between parts.

Step 3: Once all the parts are fabricated, they will meet, in a specific sequence, at the assembly line. The position of the assembly line will be determined by where the individual parts came from. At the assembly line, all flow lines join together and travel as one to packout, warehouse, and shipping.

A well-conceived flow diagram will be the best technique for developing a plant layout.

Plastic overlays of plant layouts are often used to develop flow lines for flow diagrams. The flow lines can be drawn with a grease pencil and can be grouped by classes for plants with a lot of different parts. It doesn't take a large product to make the fabrication departments of a plant layout look like a bowl of spaghetti. Using several plastic overlays will simplify the analysis.

A new industrial technologist will learn much from the creation of a flow diagram, and an experienced technologist will always find ways to improve the flow of material. Savings at this stage of methods analysis can be substantial—that is why we start with overall material flow.

NOTES ABOUT THE EXAMPLES (Figures 4-7 and 5-1)

These examples are from a gas grill manufacturing plant. The present method (Figure 4-7) shows flow starting at the bottom of the page and flowing up to stores and press shop. The proposed method (Figure 5-1) shows the receiving of the gas grill casting arriving at the back door, being stored on the trailer until needed by production. The press was moved to the back door, eliminating much travel and handling.

The flow diagram works well in conjunction with an operations chart, and we continue our example in Figures 4-8, 4-9, and 5-3.

THE OPERATIONS CHART

The operations chart (see Figure 5-2) has a circle for each operation required to fabricate each part, to assemble each to the final assembly, and to pack out the finished product. Every production step required, every job, and every part is included.

Operations charts show the introduction of raw materials at the top of the page, on a horizontal line.

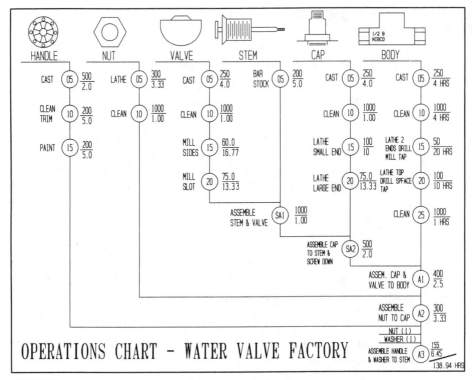

FIGURE 5-2 Operations chart for a water valve factory: complete example.

Example	Explanation
$\dfrac{1600-4250\ (2)}{\text{Stem}}$	$\dfrac{\text{Part number (How many of this part are needed)}}{\text{Part name}}$

The number of parts will determine the size and complexity of the operations chart.

Below the raw material line, a vertical line will be drawn connecting the circles (a step in the fabrication of that raw material into finished parts).

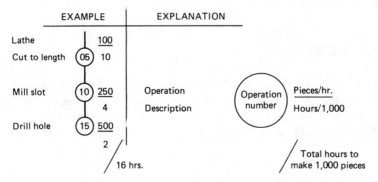

Once the fabrication steps of each part are plotted, the parts flow together in assembly. Usually, the first part to start the assembly is shown at the far right of the page. The second part is shown to the left of that, etc., working from right to left.

Example:

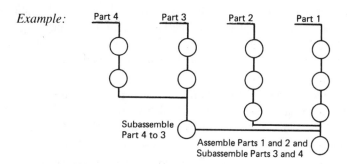

Some parts require no fabrication steps. These parts are called *buyouts*. Buyout parts are introduced above the operation at which they will be used.

Example:

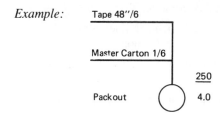

In this operation, we are going to place six products into a master carton and tape it closed.

The operations chart shows much information on one page. The raw material, the buyouts, the fabrication sequence, the assembly sequence, the equipment needs, the time standards, and even a glimpse of the plant layout, labor costs, and plant schedule can all be derived from the operations chart. Is it any wonder that industrial technologists consider this one of their favorite tools?

The operations chart is different for every product, so a standard form is not practical. The circle is universally accepted as the symbol for operations; thereby the origin of the chart's name. There is more convention in operations charting than in flow diagramming, but the designers should not be too rigid in their thinking.

STEP-BY-STEP PROCEDURES FOR PREPARING AN OPERATIONS CHART

Step 1: Identify the parts that are going to be manufactured and those that are going to be purchased complete.

Step 2: Determine the operations required to fabricate each part and the sequence of these operations.

Step 3: Determine the sequence of assembly, both buyout and fabricated parts.

Step 4: Find the base part. This is the first part that starts the assembly process. Put that part on a horizontal line at the far right top of the page. On a vertical line extending down from the right side of the horizontal line, place a circle for each operation. Beginning with the first operation, list all operations down to the last operation.

Step 5: Place the second part to the left of the first part and the third part to the left of the second part, and so forth until all manufactured parts are listed across the top of the page in reverse order of assembly. All of the fabrication steps are listed below the parts with a circle representing each operation.

Step 6: Draw a horizontal line from the bottom of the last operation of the second part to the first part just below its final fabrication operation and just above the first assembly operation. Depending on how many parts the first assembler puts together, the third, fourth, etc. parts will flow into the first part's vertical line, but always above the assembly circle for that assembly operation.

Step 7: Introduce all buyout parts on horizontal lines above the assembly operation circle where they are placed on the assembly.

Step 8: Put time standards, operation numbers, and operation descriptions next to and in the circle.

Step 9: Sum total the hours per 1,000 units and place these total hours at the bottom right under the last assembly or packout operation.

Some parts will flow together before they reach the assembly line. This could involve welding parts together or assembling a bag of parts. This is called *subassembly* and is treated just like the main assembly, except that it is done before the parts reach the vertical line on the far right top of page. Bag packing is a good example—all parts are usually buyouts and could be placed at the bottom left of your operations chart, like this:

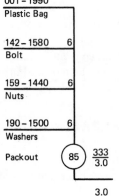

This subassembly packs out six nuts, bolts, and washers into a plastic bag.

FRED MEYERS & ASSOCIATES PROCESS CHART

| ☐ PRESENT METHOD | ☒ PROPOSED METHOD | DATE: 5/6/85 | PAGE 1 OF 1 |

PART DESCRIPTION: CASTINGS

OPERATION DESCRIPTION: PREPARE CASTING FOR PACKOUT

SUMMARY	PRESENT NO.	PRESENT TIME	PROPOSED NO.	PROPOSED TIME	DIFF. NO.	DIFF. TIME
◯ OPERATIONS	8	2452	4	1315	4	1137
⇨ TRANSPORT.	9	779	2	0	7	779
☐ INSPECTIONS	-					
D DELAYS	3				3	
▽ STORAGES	2				2	
DIST. TRAVELED	1420 FT.		240 FT.		1180 FT.	

ANALYSIS:
WHY   WHEN
WHAT   WHO
WHERE   HOW

FLOW DIAGRAM ATTACHED (IMPORTANT)

STUDIED BY: FRED MEYERS

STEP	DETAILS OF PROCESS	METHOD	OPERATION	TRANSPORT	INSPECTION	DELAY	STORAGE	DISTANCE IN FEET	QUANTITY	TIME (.00001)	COST PER UNIT	TIME/COST CALCULATIONS
1	UNLOAD	HAND	◉	⇨	☐	D	▽			31	.00217	@ $7.00/hr no cost
2	MOVE TO PUNCH	CONVEYOR	◯	⬛	☐	D	▽	40'	34.3			
3	PUNCH WINDOW HOLE		◉	⇨	☐	D	▽			642	.04494	
4	HANG		◉	⇨	☐	D	▽			321	.00225	
5	MOVE TO PAINT	OVER HEAD	◯	⬛	☐	D	▽	200'	FREE			
6	PACKOUT		◉	⇨	☐	D	▽			321	.00225	
7			◯	⇨	☐	D	▽					
8			◯	⇨	☐	D	▽					
9			◯	⇨	☐	D	▽					
10			◯	⇨	☐	D	▽				.0516	
11			◯	⇨	☐	D	▽					
12			◯	⇨	☐	D	▽					
13			◯	⇨	☐	D	▽					
14			◯	⇨	☐	D	▽					
15			◯	⇨	☐	D	▽					
16			◯	⇨	☐	D	▽					
17			◯	⇨	☐	D	▽					

FIGURE 5-3 Process chart: An improvement over Figures 4-8 and 4-9.

PROCESS CHART (Figures 4-8, 4-9 & 5-3)

The process chart (see Figure 5-3) is used to show all the handling, inspection, operations, storage, and delays that occur to one part as it moves from the receiving department through the plant to the shipping department.

Conventional symbols have been used to describe the process steps. These symbols have been accepted by every professional organization working with motion and time study.

Symbol	Description	Indicates	Meaning
◯	Circle	Operation	Performing work on a part of a product
☐	Square	Inspection	Used for quality control work
⇨	Arrow	Transportation	Used when moving material
▽	Triangle	Storage	Used for long-range storage
D	Big D	Delay	Used when storage of less than a container

Process charting lends itself to a standard form. A properly designed form will lead the designer to ask questions of each step. (The questions were asked in the introduction to this chapter, p. 44. Please review them.)

A blank form for process charting has been included at the back of this book for your future use. Examples of a completed before-and-after study are included in Figures 4-8, 4-9 and 5-3. Read and study them. Try to visualize what is happening to this part. Have all your questions been answered? The process chart communicates the same information as the flow diagram.

STEP-BY-STEP DESCRIPTION FOR USING THE PROCESS CHART

Let's look at a step-by-step example of process charting (see Figure 5-4).

①. ☐ Present Method   ☑ Proposed Method

A check mark in one of the two boxes is required. A good industrial technologist's practice is always to record the present method so the improved (proposed) method can be compared to it. Costing the present and proposed methods will be required to justify your proposal, especially if any costs are involved. Recording and advertising cost reduction dollars saved is a smart idea.

②. Date _____ Page _____ of_____

Always date your work. Our work tends to stay around for years, and you will someday want to know when you did this great work. Page numbers are important on big jobs to keep the proper order.

③. Part description

This is probably the most important information on the form. Everything else would be useless if we didn't record the part number. Each process chart is for one

FRED MEYERS & ASSOCIATES	PROCESS CHART

☐ PRESENT METHOD ①☐ PROPOSED METHOD   DATE: ② PAGE___OF___.

PART DESCRIPTION: ③

OPERATION DESCRIPTION: ④

SUMMARY	PRESENT		PROPOSED		DIFF.		ANALYSIS:		FLOW ⑦
	NO.	TIME	NO.	TIME	NO.	TIME			DIAGRAM
◯ OPERATIONS							WHY	WHEN	ATTACHED
⇨ TRANSPORT.			⑤				WHAT ⑥ WHO		(IMPORTANT)
☐ INSPECTIONS							WHERE	HOW	
D DELAYS									
▽ STORAGES							STUDIED BY:		
DIST. TRAVELED		FT.		FT.		FT.			

STEP	DETAILS OF PROCESS	METHOD	OPERATION TRANSPORT INSPECTION DELAY STORAGE	DISTANCE IN FEET	QUANTITY	TIME (0000)	DISCOUNT	COST PER UNIT	TIME/COST CALCULATIONS
1			◯ ⇨ ☐ D ▽						
2			◯ ⇨ ☐ D ▽						
3	⑧	⑨	◯ ⇨ ⑩ D ▽	⑪	⑫	⑬	⑭		⑮
4			◯ ⇨ ☐ D ▽						
5			◯ ⇨ ☐ D ▽						
6			◯ ⇨ ☐ D ▽						
7			◯ ⇨ ☐ D ▽						
8			◯ ⇨ ☐ D ▽						
9			◯ ⇨ ☐ D ▽						
10			◯ ⇨ ☐ D ▽						
11			◯ ⇨ ☐ D ▽						
12			◯ ⇨ ☐ D ▽						
13			◯ ⇨ ☐ D ▽						
14			◯ ⇨ ☐ D ▽						
15			◯ ⇨ ☐ D ▽						
16			◯ ⇨ ☐ D ▽						
17			◯ ⇨ ☐ D ▽						

FIGURE 5-4 Process chart: The step-by-step form.

part, so be specific. The part description also includes the name and specifications of the part. Attaching a blueprint to the process chart would be useful.

④. Operation description

In this block, you record the limits of the study (for example, from receiving to assembly). Also, any miscellaneous information can be placed here.

⑤. Summary

The summary is used only for the proposed solution. A count of the operations, transportation, inspection, delays, and storages for the present and proposed methods is recorded and the difference (savings) is calculated.

The distance traveled is calculated for both methods and the difference calculated. The time standards in minutes or hours is summarized and the difference calculated. This information is why we did all the work of present and proposed process charting; it is the cost reduction information. We will come back to Step 5 after Step 15.

⑥. Analysis

The questions why, what, where, when, how, and who are asked of each step (line) in the process chart, and *why* is first. If we don't have a good *why,* we can eliminate that process chart step and save 100% of the cost. Questioning each step is how we come up with the proposed method. With these questions, we try to

1. Eliminate every step possible, because this produces the greatest savings. However, if we can't eliminate the step, we try to

2. Combine steps to spread the cost and possibly eliminate steps between. For example, if two operations are combined, delays and transportations can be eliminated. If transportations are combined, many parts will be handled as one. However, if we can't eliminate or combine, maybe we can

3. Change the sequence of operations to improve the product flow and save many feet of travel.

As you can see, the analysis phase of process charting gives the process meaning and purpose. We will come back to Step 6 after Step 15.

⑦. Flow diagram attached (important)

Process charting is used in conjunction with flow diagramming. The same symbols can be used in both techniques. The process chart is the words and numbers, whereas the flow diagram is the picture. The present and proposed methods of both techniques must be telling the same story; they must agree.

Your name goes in the section labeled "Studied By." Be proud and put your signature here.

⑧. Details of process

Each line in the flow process chart is numbered, front and back. One chart can be used for forty-two steps. Each step is totally independent and stands alone. A description of what happens in each step aids the analyst's questions. Using as few words as possible, describe what is happening. This column is never left blank.

⑨. Method

Method usually refers to how the material was transported—fork truck, hand cart, conveyer, hand, etc.—but methods of storage could also be placed here.

⑩. Symbols

The process chart symbols are all here. The analyst should classify each step and shade the proper symbol to indicate to everyone what this step is.

⑪. Distance in feet

This step is used only with the transportation symbol. The sum of this column is the distance traveled in this method. This column is one of the best indications of productivity.

⑫. Quantity

Quantity refers to many things:

a. Operations: The pieces per hour would be recorded here.

b. Transportation: How many were moved at a time

c. Inspection: How many pieces per hour if under time standard and/or frequency of inspection

d. Delays: How many pieces in a container. This will tell us how long the delay is.

e. Storage: How many pieces per storage unit.

All costs will be reduced to a unit cost or cost per 1,000 units, so knowing how many pieces are moved at one time is important.

⑬. Time in hours per unit (.00001)

This step is for labor costing. Storage and delays will be costed in another way: inventory carrying cost. This column will be used only for operations, transportations, and inspection. Time per unit is calculated in two ways:

a. Starting with pieces per hour time standards, say 250 pieces per hour, divide 250 pieces per hour into 1 hour, and you get .00400 hours per unit. On our process chart, we place 400 in the time column, knowing that the decimal is always in the fifth place.

b. The material handling time to change a tub of parts at a work station with a hand truck is 1.000 minutes, and we have 200 parts in that tub. How many hours per unit is our time standard?

$$\frac{1.000 \text{ min./container}}{200 \text{ parts/container}} = .005 \text{ min./part}$$

$$\frac{.005 \text{ min./part}}{60 \text{ min./hr.}} = .00008 \text{ hr./part}$$

⑭. Cost per unit

Hours per unit multiplied by a labor rate per hour equals a cost per unit. For example, consider the aforementioned two problems using a labor rate of $7.50 per hour:

a. $.00400 \times 7.50 = \$ .03$ per unit

b. $.00008 \times 7.5 = \$ .0006$ per unit.

The cost per unit is the backbone of process charting. We are looking for a better way, so the method with the overall cheapest way is the best method.

⑮. Time/cost calculations

Technologists are required to calculate costs on many different things, and how costs were calculated tends to get lost. This space is provided to record the formulas developed to determine the costs so they do not have to be redeveloped over and over again.

⑤. Returning to *Summary*

Once all steps in the present method process chart have been completed, the summary is completed by

a. Counting all the operations, transportations, etc.

b. Adding up the unit time for all steps

c. Adding up the distance traveled.

⑥. Returning to *Analysis*

Once the present method is recorded and costed, the search for a better method starts. It is not appropriate to start the analysis phase until after the completed present method has been recorded. During the analysis phase, a proposed method is being developed and the industrial technologist goes through the process all over again. Figure 5-3 is an improved process chart for the gas grill housing in Figures 4-8 and 4-9. How much was saved?

Do you remember what was said in an earlier chapter about successful people doing what others do not like to do? Working hard was one of those things. There is no easy way to finding the best method and selling that method to management—only hard work using the tried-and-true techniques of motion and time study. The process chart is one of the best techniques and one of the most widely used.

FLOW PROCESS CHART

The flow process chart (see Figure 5-5) combines the operations chart with the process chart. The operations chart used only one symbol—the circle or operations symbol. The flow process chart is just five times more, using all five process chart symbols. Another difference is that buyout parts are treated like manufactured parts. No standard form exists for flow process charting.

The flow process chart is the most complete of all the techniques, and when completed the technologist will know more about the plant's operation than anyone in that plant.

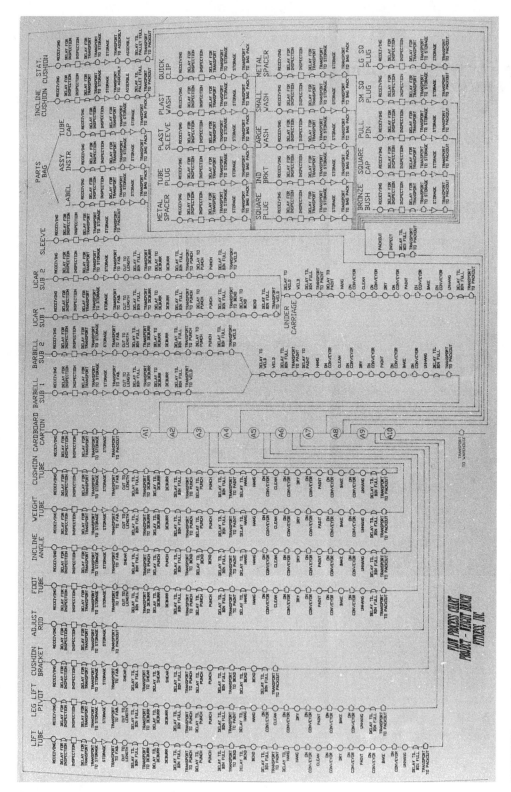

FIGURE 5-5 Flow process chart: Shows flow of every part through the plant.

STEP-BY-STEP PROCEDURE FOR PREPARING A
FLOW PROCESS CHART

Step 1: Start with an operations chart.

Step 2: Complete a process chart for each part.

Step 3: Combine the operations chart and the process chart working in all the buyouts.

CONCLUSION

The broad view of motion study looks at the entire operation first because

1. The overview identifies all the steps required to manufacture a product and give the technologist a game plan.
2. Cost savings at the overview level tend to be larger than savings at the individual operations level.
3. The overview is needed for bigger questions such as plant layout, cost estimating, and scheduling.

Once the broad view methods design is complete, the industrial technologist will focus on individual operations and improve them.

QUESTIONS

1. Why do we start our methods study with the broad view?
2. Describe flow diagrams, operations charts, process charts, and flow process charts. When would you use each technique?
3. What is the most important reason for using any of these techniques?
4. Describe the use of the cost reduction formula.
5. What is cross traffic and backtracking? What is our attitude toward these?
6. Why is distance traveled important?
7. What are the five process chart symbols? Give examples.
8. Develop a flow process chart for the water value plant shown in Figure 5-2.

CHAPTER 6

Techniques of Motion Study: Operations Analysis

INTRODUCTION

The techniques of the previous chapter identified the operations required to produce a product. This was called the broad view. In this chapter, the techniques of motion study of individual operations are discussed. The purposes of these techniques are to understand the operations of these work stations and to improve them. This chapter and the next two concentrates on work station design, also known as work design and work simplification. The techniques of operations analysis, motion economy theory, and predetermined time standards systems (PTSS) will allow us to design and cost the most efficient method of production. Our goal is to produce a product with as little effort and cost as possible. The limits of cost reduction are only set in our minds.

The techniques of motion study discussed in this chapter are

1. Operations analysis charting
2. Operator/machine charting
3. Gang charting
4. Multimachine charting
5. Left-hand/right-hand charting.

These techniques have several things in common:

1. Each activity is broken up into elements. An activity is one unit of production. For example, if one operator were to operate three pieces of machinery, there would be four activities, one operator, and three machines. Under the operator activity, we may

have several elements of work for each machine, while the machine activity would have two elements—working or idle.

2. Time is measured linearly. A scale is drawn down the side of an activity, and the unit of measurement is in minutes, usually .01 (one hundredth) of a minute. The elements would be divided by a horizontal line indicating how much time is required.

3. All operations analysis techniques can use the same form. Only the number of activities varies. In the foregoing four-activity example, we would need two of the standard forms side by side. The operations analysis chart would use only one half of one form.

4. All of these charting techniques are visual and are good sales tools. The length of the operation is shown by a scale on the chart, and the present method can be held next to the proposed method, with the choice being obvious.

OPERATIONS ANALYSIS CHART

The operations analysis chart (see Figure 6-1) is used to describe a single activity, usually one operator using only tools and equipment that are totally operator controlled. This is the simplest of all the charts discussed in this chapter, because it has only the one activity; however, the process is the same process used for the most difficult chart. The single activity is broken up into elements (an element of work is one unit of work that cannot be divided realistically), and these elements are timed. The method of timing is not important at this stage, but decimal minutes are used.

Example: A packout operator on an assembly line is required to pack the following parts into a carton passing on the assembly line:

Part Number	Quantity/Set	Description	Time
1	2	Leg	.150
2	1	Bar	.065
3	2	Seat	.125

The assembly line runs at .400 minutes/set. The 100% station on the assembly line (normally only one station) is fully loaded and will have .400 minutes of work.

The operations analysis chart will show the aforementioned information on a vertical time scale, with the size of each element in direct proportion to the amount of time that element takes.

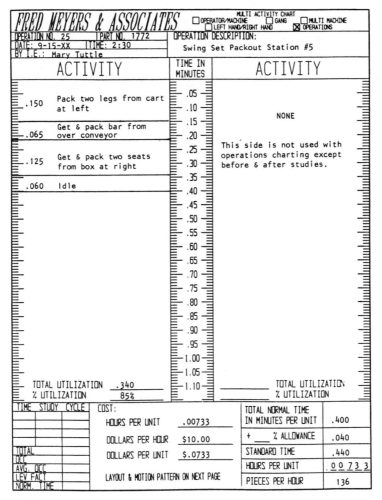

FIGURE 6-1 Operations analysis chart.

OPERATOR/MACHINE CHART

The operator/machine chart (see Figures 6-2 and 6-3) is twice as complicated as the operations analysis chart. The operator/machine chart has two activities—the operator and the machine. The operator/machine chart is much more useful because it shows the interrelationship of the operator and the machine. Both the machine and the operator work intermittently, and this chart shows what each is doing at every moment in time. Each activity is reduced to a series of elements (an element is that unit of work that cannot be subdivided), and these elements of work are placed in order down one side

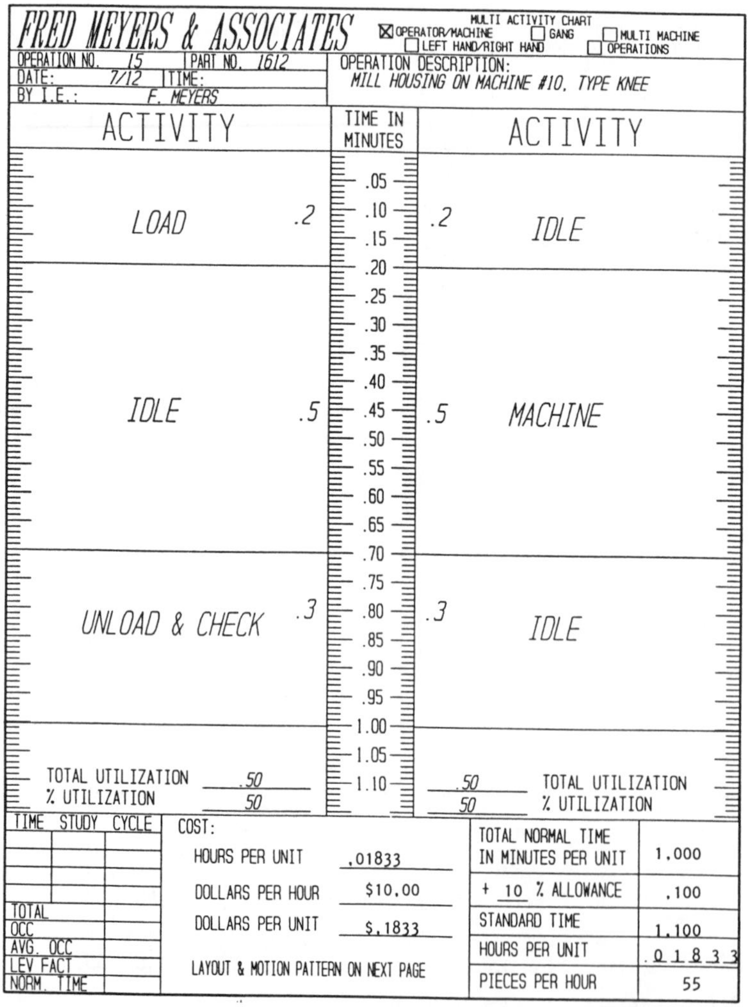

FIGURE 6-2 Operator/machine chart: Two activities working together.

of the chart while the other activity's elements are placed down the opposite side of the chart. Each element must be aligned, time wise, so the same moments are across from each other.

NOTES ON FIGURE 6-2 (PRESENT METHOD)

Activity 1 is the operator (left side), and Activity 2 is the machine (right side). The first element of the operator is the load element, and the first element of the machine is an idle element. These two elements end when the operator pushes the start button to

FRED MEYERS & ASSOCIATES		MULTI ACTIVITY CHART		
		☒ OPERATOR/MACHINE ☐ GANG ☐ MULTI MACHINE		
		☐ LEFT HAND/RIGHT HAND ☐ OPERATIONS		

OPERATION NO. _5_	PART NO. _1612_	OPERATION DESCRIPTION:
DATE: _7/12_	TIME:	_MILL VALVE BODY ON MACHINE #16_
BY I.E.: _F. MEYERS_		

ACTIVITY	TIME IN MINUTES	ACTIVITY
LOAD	.05	IDLE
	.10	
PRE POSITION & ASIDE	.15	
	.20	
	.25	MACHINE TIME
	.30	(MILL SURFACE)
	.35	
	.40	
	.45	
	.50	
IDLE	.55	
	.60	
UNLOAD	.65	IDLE
	.70	
	.75	
	.80	
	.85	
	.90	
	.95	
	1.00	
	1.05	
TOTAL UTILIZATION ___.60___	1.10	___.50___ TOTAL UTILIZATION
% UTILIZATION ___86___		___71___ % UTILIZATION

TIME STUDY CYCLE			COST:		TOTAL NORMAL TIME IN MINUTES PER UNIT	.700
.75	3.80	6.80	HOURS PER UNIT	_.01295_		
1.55	4.50	7.50			+ _11_ % ALLOWANCE	.077
2.20	5.25		DOLLARS PER HOUR	$10.00	STANDARD TIME	.777
3.05	6.05		DOLLARS PER UNIT	$ .1295		
TOTAL	7.50				HOURS PER UNIT	.01295
OCC	10					
AVG. OCC	.750		LAYOUT & MOTION PATTERN ON NEXT PAGE		PIECES PER HOUR	77
LEV FACT	95%					
NORM. TIME	.712					

FIGURE 6-3 Operator/machine chart: An improvement over Figure 6-2.

activate the machine. At this moment, the operator starts the idle element while the machine starts its work element. When the machine stops cutting, the machine work element ends; at the same moment, the operator's idle element ends. This also starts the last elements: The operator unloads the machine and asides the part, and the machine is again idle. These elements end when the operator puts down the finished part and starts reaching for the next part, which starts the whole process again. Both activities took 1.00 minutes, and both activities actually worked only 50% of the time. Is this good utilization of our company's resources—people and machinery?

NOTES ON FIGURE 6-3 (PROPOSED METHOD)

This is the same job as in Figure 6-2, but during the machine element of the machine activity, the operator prepositioned and asided the finished part. Hold the present method and the proposed method side by side. See the improvement? Any doubt which is the best procedure? If one million of these parts are produced each year, what is the savings?

Present method	$.1833 per unit	.1833
Proposed method	$.1295 per unit	.1295
Savings	$.0538 per unit	.0538
Times 1,000,000 units per year equals		
	$53,800/per year	

STEP-BY-STEP PROCEDURE FOR PRODUCING ALL THE CHARTS IN THIS CHAPTER

See Figure 6-4 for an example of a multiactivity chart.

Step ①: Identify the problem: Operations analysis, operator/machine; gang, multimachine, left-hand/right-hand chart. The only differences are the number of activities and the kind of activities. The examples given in this chapter will help identify which technique to use for each task. Check the appropriate box once the problem has been identified.

Step ②: Operation number: The operations required to manufacture every part are numbered to identify them. The operations numbers can be anything the systems designer wants, but usually the first operation number is 05 followed by 10, 15, 20, etc. A few parts would require more than twenty operations, but for those parts, operation numbers between the one divisible by 5 may be used. The system of skipping numbers gives room for expanding the operations if a new operation is required between 15 and 20, so this new operation could be 17 or 18—your choice.

Step ③: Part number: Every product is made up of parts. To keep control of these parts, each part is assigned a part number. There are many techniques for setting up part numbers, and most companies try to make part numbers meaningful and useful. One useful system is an eight-digit system—XXXX = XXXX. The first four digits identify the finished product, while the last four digits identify the part. In addition, the last four digits can be very meaningful by identifying parts groups, such as

9XXX   Packaging

8XXX   Plastics

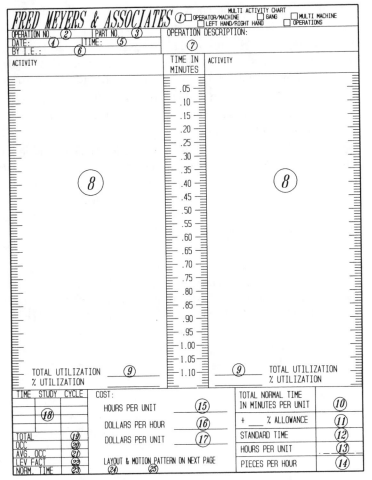

FIGURE 6-4 Multiactivity chart: The step-by-step form

7XXX	Paints
6XXX	Sheet metal parts
5XXX	Castings
	Etc.

Any industrial technology work done on any part should carry a part number and operation number, or this work will be useless. After a short period of time, the busy technologist will not remember what the work was for.

Step ④*:* Date: Always include the date on your work. This includes the year. Your work will stay around for years, and subsequent studies will be compared to older ones. How will you know the sequence of studies unless you have the date?

Step ⑤: Time: Time may affect the study. For example, a machine or process may work better or worse in the morning before it warms up. You may not know this at the time of your study, so just in case, record the time of the study.

Step ⑥: "By I.E.": The I.E. refers to industrial engineer, your professional service. Your name as the analyst, technologist, or engineer goes in this space. A study without a name is worthless.

Step ⑦: Operation description: The operation description should be as complete as possible. You must be able to communicate what is happening in such detail that the next generation of engineers can understand what you did. Every senior industrial technologist has experienced a time when he or she couldn't figure out what was done six months ago for lack of an operation description. Don't save time by being sloppy and incomplete; do it right the first time—do it now.

Step ⑧: Activity name: The activity is the guts of the chart. The activity name is the operator's job description name or machine name. Under each activity name, the elements of the job are listed in the exact order they are performed. The size of the elements are dependent on the time required. The more time required, the larger the element size. The time is measured on the linear scale at the side of each activity. In the center is the cumulative time. Time can be set by any of the techniques studied later in this book, but do not worry about this now.

Step ⑨: Total utilization, % utilization: This comprises two pieces of information. First, total utilization is how much time this activity was working. This time does not include idle time. The second piece of information is percent utilization, which indicates how well the activity (operator or machine) is being used. The percent utilization is calculated by dividing the total utilization of the activity by the total time of that activity. In Figure 6-2, for example,

$$100 \times \frac{.5 \text{ minutes total utilization}}{1.0 \text{ minutes total time}} = 50\% \text{ utilization.}$$

Step ⑩: Total normal time in minutes per unit: Total normal time is how long it takes each activity to complete a cycle and includes both working and idle time. The total normal time will be the same for all activities of an operation. In Figure 6-2 of the operator/machine chart, the total normal time was 1.00 minutes.

Step ⑪: +__% allowance: Allowances are the time we add to normal time to make our time standards realistic. Allowances include time for personal time, fatigue (breaks), and unavoidable delays. We will study allowances in the time study chapter (Chapter 9), but until then, we will use 10% allowances because it is the most common. So, we have +10% allowance. 10% of the total normal time in minutes per unit is placed here. In our example, 10% of 1.00 minutes is .10 minutes.

Step ⑫: Standard time: Standard time is normal time plus allowances. The difference between normal time and standard time is allowances.

Step ⑬: Hours per unit: Hours per unit is calculated by dividing standard time by 60 minutes/hour. In our example,

$$\frac{1.10 \text{ minutes/unit}}{60 \text{ minutes/hour}} = .01833 \text{ hours/unit.}$$

Multiply .01833 hours per unit by 1,000 units to determine the hours per 1,000 units or 18.33 hours/1,000. Hours per 1,000 is a more meaningful number than hours per unit.

.01833   hours per unit

or

18.33   hours per 1,000.

Which is the most meaningful to you?

Step ⑭: Pieces per hour: Pieces per hour is calculated by dividing the hours per unit into 1 or the hours/1,000 into 1,000. In our example,

$$\frac{1}{.01833} = 55 \text{ pieces/hour} \quad \frac{1,000}{18.33} = 55 \text{ pieces/hour.}$$

Step ⑮: Hours per unit: This is the same number calculated in Step 13 and is the first ingredient of cost.

Step ⑯: Dollars per hour: The dollars per hour refers to the operator's wage rate in $/hr. This can be more complicated than shown in Figure 6-3, but a useful method is the departmental average labor rate plus fringe benefit rate. All the employees' salaries of a department are added together and divided by the number of employees to get the average hourly rate. The cost of employee benefits, such as vacation, holiday, and insurance, is calculated yearly and converted to a percentage of the hourly rate. 33.3% is a common fringe number, so if our hourly average rate is $7.50/hour, fringe benefits (7.50 × 33.3%) equals $2.50, for a total cost of $10.00 per hour.

Step ⑰: Dollars per unit: The dollars per unit is the labor cost for one unit. To calculate the dollars per unit, multiply (Step 15 × Step 16) hours per unit times dollars per hour. This can be complicated by more than one operator; in that case, cost would double. Dollars per unit is our measure of desirability. We want this to be as low as possible.

Step ⑱: Time study cycles: The time study cycles block is a small time study form and allows the analyst a place to check his or her work. There are twelve blocks in which to place stopwatch time study observation readings. When the operator finishes a unit, the industrial technologist will start the stopwatch and record the ending time of the next twelve parts. The watch could be left running (continuous time study) or reset each time a part is run (snap-back time study). These two techniques are discussed in the time study chapter, chapter 9.

Step ⑲: Total: On a continuous time study, the total is the last reading. The last reading would be the total time for running twelve parts. On the snap-back time study, each reading is the time for one part; therefore, the technician needs to add the twelve readings together to get the total time.

Step ⑳: Occurrences: This is the number of occurrences or cycles. If the technician studied all twelve cycles, then this figure is 12. But if the time study technician studied less than twelve cycles, the number of cycles studied goes in this block.

Step ㉑: Average occurrence: The average occurrence is the average time per part. It is calculated by dividing the total time (Step 19) by the occurrence (Step 20). Average time per occurrence is simply the arithmetic average of the cycles checked.

Step ㉒: Leveling factor: The leveling factor is a complicated subject that is discussed fully in Chapter 9. At this time, the best definition is that the leveling factor is the time study technician's opinion of the speed or tempo of the operator, or, in simpler terms, how fast the operator is working. Normal is 100% performance; therefore, if someone is working less than normal, we would put a leveling factor of less than 100. An operator working faster than normal would be leveled at over 100%. (An example is shown in Step 23.)

Step ㉓: Normal Time: Normal time is calculated by multiplying the average occurrence time (Step 21) by leveling factor (Step 22).

Examples:

Average Occurrence	Leveling Factor	Normal Time in Minutes
1.00	120%	1.20
1.00	80%	.80
1.00	100%	1.00

Note that the leveling factor makes a big difference in normal time; therefore, a time study analyst needs much practice in leveling. The normal time in Step 23 needs to be compared to the total normal time in Step 10. If they are close, the analyst will feel confident that this time standard is a good one.

Step ㉔: Layout: A top-view drawing of the work station layout is very important to a complete description of the operation (Step 7). Chapter 7 is on work station layout, so further discussion is postponed until then.

Step ㉕: Motion pattern: The motion pattern is the path made by the hands in the process of making one part. The motion pattern is the modern technique which replaces the cyclograph shown in chapter 2. The motion pattern is also a major discussion of chapter 7, so further discussion is postponed until then.

The operator machine chart is the most popular of the multiactivity charts, but all charts are similar. Only deviations from the aforementioned procedures are discussed in the future charts and will be titled "Procedure Deviations."

GANG CHART

A gang chart (see Figure 6-5) is used when two or more people are working together and their activities intertwine. A two-person team gang chart would look just like an operator/machine chart. When three or more operators are involved, more columns are needed on the standard form. This is accomplished by taping two or more pages side by side, building as many columns as needed, as shown in Figure 6-5. All multiactivity charts have idle time. Our goal is to minimize this idle time and to ensure proper crew size.

Procedure Deviations

1. Hours per unit will be multiples of one operator's time. Take the number of operators working together times the hours per unit for one person.
2. Pieces per hour will still be calculated from the one person hours per unit, just the same as the operator/machine chart.
3. Cost must reflect all the crew size, so hours per unit must reflect all the operators.

NOTES ABOUT FIGURE 6-5

Swing sets are packed out in boxes 10' long, 18" wide, and 8" high and weigh 125 pounds. Finished swing sets must be removed from the packout line continually, stacked on hand carts (using two people because of weight and length), rolled to a storage location, and restacked on the floor to a 12' height. The activities are

1. Two people remove sets and stack on carts.
2. Carts are rolled to stacks and an empty cart brought back.
3. Two people remove sets from roller cart and hand to a stacker.
4. Person on top of stack straightens sets to align and cross tie each tier.

The big question is, how many cart roller operators do we need?

EXAMPLE: GANG CHART

THESE PEOPLE ARE A TEAM. IS THERE ANYWAY TO

IMPROVE THE BALANCE OF WORK?

FRED MEYERS & ASSOCIATES

MULTI ACTIVITY CHART
☐ OPERATOR/MACHINE ☒ GANG ☐ MULTI MACHINE
☐ LEFT HAND/RIGHT HAND ☐ OPERATIONS

DIVISION NO. 99 PART NO. 1772
DATE: 9-15-XX TIME: NOON
BY J.E.: MEYERS

OPERATION DESCRIPTION:
UNLOAD FINISHED SWING SETS FROM PACK
OUT LINE & STACK IN WAREHOUSE

2 PEOPLE	ACTIVITY	REMOVE FROM CONVEYOR	TIME IN MINUTES	1 PERSON	ACTIVITY	MOVE CARTS
.14	REMOVE FINISHED SET FROM LINE & STACK #1 ON CART (10/CART)		.05 / .10 / .15		REMOVE FULL CART PLACE EMPTY CART	.15
.06	IDLE		.20 / .25 / .30			
.14	REMOVE & STACK #2		.35 / .40			
.06	IDLE		.45 / .50		MOVE FULL CART TO TO WAREHOUSE 10/CART	.8
.14	REMOVE & STACK #3		.55 / .60			
.06	IDLE		.65 / .70 / .75			
	REMOVE & STACK #4		.80 / .85			
	IDLE		.90 / .95 / 1.00			
	REMOVE & STACK #5		.05 / .10			
	IDLE		.15 / .20			
.14	REMOVE & STACK #6		.25 / .30		BRING EMPTY CART BACK	
.06	IDLE		.35 / .40			
.14	REMOVE & STACK #7		.45 / .50			
.06	IDLE		.55 / .60			
.14	REMOVE & STACK #8		.65 / .70			
.06	IDLE		.75 / .80			
.14	REMOVE & STACK #9		.85 / .90		IDLE	.19
.06	IDLE		.95 / 2.00			
.14	REMOVE & STACK #10		2.05		REMOVE FULL CART CONTINUED	.06
.06	IDLE		2.10			
.1	TOTAL UTILIZATION 1.40 % UTILIZATION 70				TOTAL UTILIZATION 1.71 % UTILIZATION 90.5	

2 PEOPLE	ACTIVITY	REMOVE FROM CONVEYOR	TIME IN MINUTES	1 PERSON	ACTIVITY	MOVE CARTS
.03	STACK 6 CONT.		.05 / .10		STRAIGHTEN #6	.03
.18	STACK 7		.15 / .20		IDLE	.08
.18	ON FLOOR CROSS TIE 7 PER TIER		.25 / .30		STRAIGHTEN 7	.10
					IDLE	.08
.18	STACK 8		.35 / .40		STRAIGHTEN 8	.10
.18	STACK 9		.45 / .50		IDLE	.08
			.55 / .60		STRAIGHTEN 9	.10
.18	STACK 10		.65 / .70 / .75		IDLE	.08
					STRAIGHTEN 10	.10
.20	IDLE		.80 / .85 / .90 / .95		IDLE	.28
.18	STACK 1		1.00 / .05 / .10		STRAIGHTEN 1	.10
.18	STACK 2		.15 / .20 / .25 / .30		IDLE STRAIGHTEN 2	.08 / .10
.18	STACK 3		.35 / .40 / .45		IDLE STRAIGHTEN 3	.08 / .10
	STACK 4		.50 / .55 / .60		IDLE STRAIGHEN 4	.08 / .10
	STACK 5		.65 / .70 / .75 / .80		IDLE STRAIGHTEN 5	.08 / .10
	STACK 6		.85 / .90 / .95		IDLE	.08
	CONTINUED		2.00 / 2.05		STRAIGHTEN 6 CONTINUED	.07
	TOTAL UTILIZATION 1.80 % UTILIZATION 90%		2.10		TOTAL UTILIZATION 1.71 % UTILIZATION 90.5	

NOTES ON COST & STANDARD:
1. 273 SWING SETS MUST BE REMOVED FROM THE LINE PER HOUR BECAUSE THE ASSEMBLY LINE IS MOVING THAT FAST.
2. THE HOURS PER UNIT MUST BE MULTIPLIED BY 6 OPERATORS.

COST:

HOURS PER UNIT	.022		TOTAL NORMAL TIME IN MINUTES PER UNIT	.200
DOLLARS PER HOUR	$10.00		+ 10 % ALLOWANCE	.020
DOLLARS PER UNIT	* .22		STANDARD TIME	.220
* FOR ALL 6 PEOPLE LAYOUT & MOTION PATTERN ON NEXT PAGE			HOURS PER UNIT	.02200
			PIECES PER HOUR	273

FIGURE 6-5 Example: Gang chart. These people are a team. Is there any way to improve the balance of work?

Time Study Information

1. Sets are coming out of production at a rate of five per minute. The operators are busy 70% of the time. Sets are stacked ten per cart. Ten sets weigh 1,250 pounds plus cart weight.

2. A cart can be rolled to the warehouse and back in 1.6 minutes.

3. Sets can be stacked in .18 minutes per set or faster, depending on stack height.

4. Once boxes are placed on stack, a third person straightens and and aligns them in .10 minutes per box.

MULTIMACHINE CHART

When an operator is asked to run more than one machine, the multimachine chart is used (see Figure 6-6). How many machines can one person operate? The multimachine chart can show us. The operator/machine chart is expanded so each machine has a column of its own. The longer the cycle, the more machines a person can operate. The multimachine chart is very similar to the gang chart in appearance. In our example (Figure 6-6), how many machines could our operator run?

Procedure Deviations (Steps 2 and 3)

1. More than one part and one operation are possible. Be sure to include all part numbers and operation numbers.

2. Total normal time (Step 10) will be the same for each activity. Total normal time is the operator's time, and when an operator runs three machines, that time must be

FIGURE 6-6 Example: Multimachine chart. All three machines are the same and are doing the same job. Different parts/machines are possible. How can this method be improved?

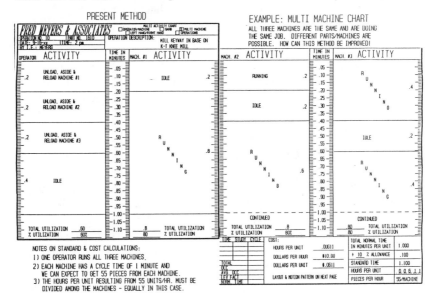

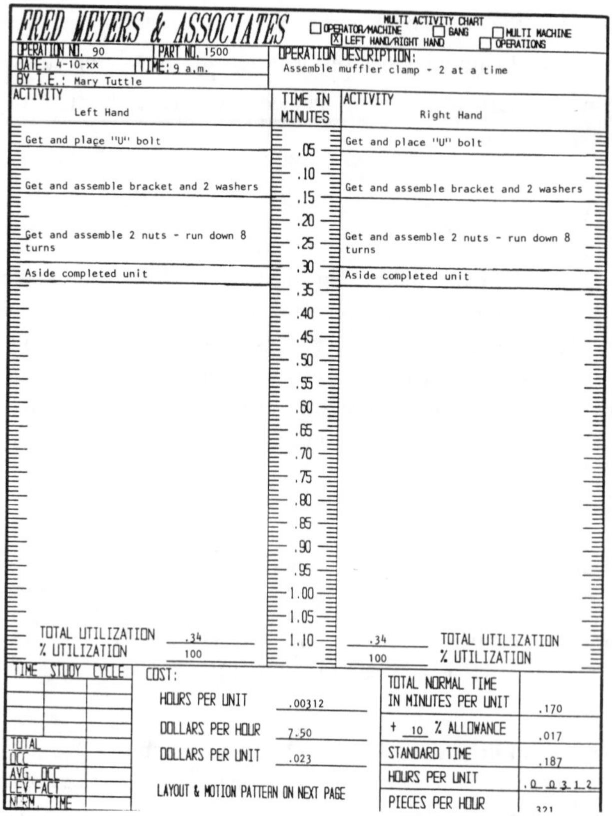

FIGURE 6-7 Example left hand/right hand chart for assembly of two muffler clamps

split among the machines in proportion to the cycle times of the machines. This is done by dividing the hours per unit (Step 13) by the number of machines. The hours per unit can be multiplied by the percentage of the operator's productive time required for each machine.

LEFT-HAND/RIGHT-HAND CHART

The left-hand/right-hand chart (see Figure 6-7) is very different from the previous charts because it is for one operator only. It is also different from the operations chart because it treats each hand as an activity. Each hand's activity is broken up into elements and is plotted in the column adjacent to the other hand, each moment being exactly across from the other. This chart is useful in showing idle time by either hand. When one hand is idle or being used as a fixture (just holding a part), this time is shown as blank space and is referred to as a "one-arm bandit." One-hand operations are inefficient and must

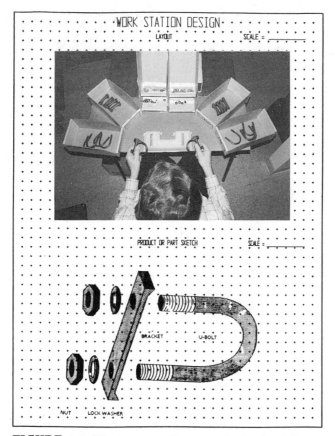

FIGURE 6-8 Work station design is the back of the time study form.

be eliminated. The left-hand/right-hand chart will give utilization rates on both hands. Full (100%) utilization of both hands is impossible, but by being creative a technician can approach full utilization. There are no procedural deviations from the operator/ machine chart standard procedure.

Example: Assemble U Bolt

Figure 6-8 is a picture of the work station and a drawing of the product. Figure 6-8 is back of the Figure 6-7. This picture and drawing help the technologist describe the operation.

QUESTIONS

1. Which is the most common use of the multiactivity chart?

2. Why don't we use the left-hand/right-hand chart more?

3. Review the cost reduction calculation for Figures 6-2 and 6-3 for 1,000,000 units per year.

4. What is an activity? How many activities are there when two operators operate five machines?

5. When one operator runs five machines each producing 200 pieces per hour, what are the standards?
   ____pieces/hour
   ____hours/piece

6. When ten people work together on a gang job and they produce 500 units per hour and earn $10.00 per hour each, what is the cost of one unit?

7. Do a left-hand/right-hand chart of your project. Compare this to your PTSS after chapter 8.

CHAPTER 7

Work Station Design

Many of the previous techniques required work station designs. The work station design is a drawing, normally top view, of the work station including equipment, materials, and operator space. The design of work stations has been an activity for industrial engineers and technologists for nearly a century. During this time, the profession has developed a list of principles of motion economy that all new technologists should learn and apply. When the principles of motion economy are properly applied to the design of a work station, the most efficient motion pattern results.

This chapter is divided into three sections:

Work station design
Principles of motion economy
Motion patterns.

WORK STATION DESIGN

The first question new technologists ask is where to start. The answer is very simple—anywhere. No matter where you start in work station design, another idea will come along and make the starting point obsolete. Where to start depends a great deal on what is to be accomplished at the work station. The cheapest way to get into production is the best rule for the starting point. The cheapest way means just that—the simplest machines, equipment, and work stations. Any improvement on this cheapest method must be justified by savings. Therefore, the technologist is free to start anywhere, and then improve on the first method.

Our first work station example is a simple assembly station for the muffler clamp

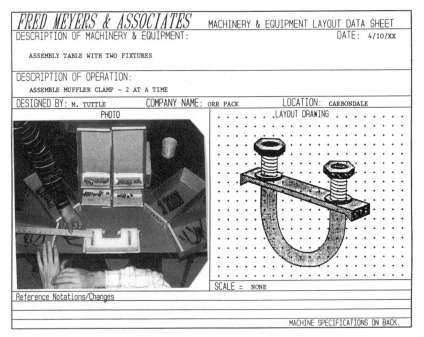

FIGURE 7-1 Muffler clamp assembly.

assembly shown in Figure 7-1. A work table of $3' \times 3' \times 42''$ high with the parts placed on top would be the simplest equipment.

The following information must be included in the design:

1. Work table
2. Incoming materials (belt, clamp, washers, and nuts; packaging and quantity must be considered)
3. Outgoing material (finished product)
4. Operator space and access to equipment
5. Location of waste and rejects
6. Fixture and tools
7. Scale of drawing.

A three-dimensional drawing would show a great deal more information, and a talented technician would attempt this.

Costing and improving this work station design is accomplished in chapter 8, where we cost and improve this job.

Our second example of work station design is a machine operation (see Figure 7-2). The needs of the station design are the same as in the foregoing list, but the equipment (machines, jigs, and fixture) will be added.

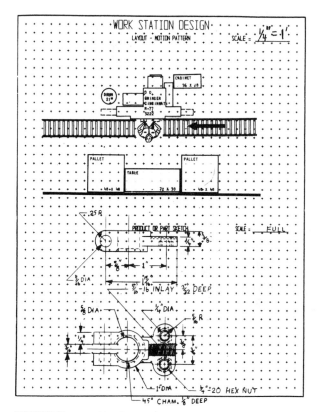

FIGURE 7-2 Machine operation.

PRINCIPLES OF MOTION ECONOMY

Industrial engineers and technologists have been continually developing guidelines for efficient and effective work station design. These guidelines have been collected and titled the Principles of Motion Economy. These principles have become a part of every motion & time study book.

Effectiveness is doing right things (the job), and efficiency is doing things right (method). So effectiveness and efficiency mean doing right things right. Effectiveness is important to consider first because doing a job that is not necessary is bad, but making a useless job efficient is the worst sin. Efficiency or doing things right is the goal of industrial technology.

The principles of motion economy should be considered for every job. Sometimes principles will be violated with good reasons. These violations and reasons should be written up for future use. You will have to defend yourself to every new technologist—be prepared.

The principles are often used together in creative ways. The only limit to improved work station design is the technologist's creativity.

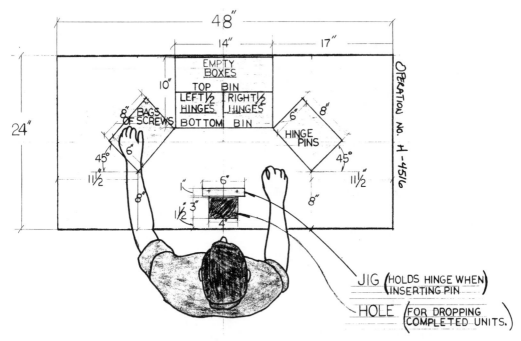

FIGURE 7-3 Design work stations to promote mirror image motion patterns.

Hand Motions

The hands should operate as mirror images (see Figure 7-3). They should start and stop motions at the same time, they should move in opposite directions, and they should both be working at all times.

If the hands are reaching for two parts at the same time, the bins should be placed equally away from the work area and the same distance from the center line of the work station. To design a normal work station, the technologist should place all parts and tools between the normal and maximum reach, but make the reaches as short as possible (see Figure 7-4).

Reaching for only one part leaves the question of what the other hand is going to do. To keep both hands working at all times is a challenge and can be most easily accomplished by doing two parts at a time (one with the left hand and one with the right). Holding parts in one hand while assembling other parts to it is a poor use of the holding hand. It is said that the most expensive fixture in the world is the human hand (see Figure 7-5). In work station design, we don't consider people righthanded or left-handed, unless hand tools are used. Then we consider everyone righthanded.

FIGURE 7-4 Normal reach versus maximum reach.

FIGURE 7-5 The most expensive fixture: The human hand.

Basic Motion Types

Ballistic Motions are fast motions created by putting one set of muscles in motion and not trying to end those motions by using other muscles. Throwing a part in a tub and hitting a panic button on a machine are good examples. Ballistic motions should be encouraged.

Controlled or Restricted Motions are the opposite of ballistic motions and require more control, especially at the end of the motion. Placing parts carefully is an example of a controlled motion. Safety and quality considerations are the best justification for controlled motions. But if ways of substituting ballistic motions for controlled motions can be found, cost reduction can result.

Continuous Motions are curved motions and more natural. When the body has to change direction, speed is reduced, and two separate motions result. If direction is changed less than 120 degrees, two motions are required. Reaching into a box of parts lying flat on the table is an example requiring two motions, one motion to the lip of the box, and another down into the box. If the box were placed on an angle, one motion could be used. (We look at this principle again when we discuss gravity later in this chapter.)

Location of Parts and Tools

Have a fixed place for everything and have everything as close to the point of use as possible. Having a fixed place for all parts and tools aids in habit forming and speeds up the learning process. Have you ever needed a pair of scissors, and when you looked where they were supposed to be, they were gone? How efficient were you in the next few minutes? A tool maker's tool box is arranged so the tool maker knows where every tool is and can retrieve it without looking. That should be our goal in every work station we design.

The need for placing parts as close to the point of use as possible will be evident in the next chapter. However, for now we can just say that the farther you reach for something, the more costly that reach will be. Creativity is required to minimize reaches. We can tier parts; instead of having one row of parts across the top of the work station, maybe three rows of parts one over the other would be better (see Figure 2-2). We can hang tools from counterbalances over the work station (see Figure 7-6). We can use conveyers to move parts into and out of the work station.

Release the Hands of as Much Work as Possible

As stated earlier, the hand is the most expensive fixture a designer can use. So we must provide other means of holding parts. Fixtures and jigs are designed to hold parts so operators can use both hands (see Figure 7-7). Foot-operated control devices can be designed to operate equipment to relieve hands of work. Conveyers can move parts past operators, so operators don't have to get or aside the base unit (see Figure 7-8). Powered round tables are also used to move parts past an operator.

Fixtures can be electric, air, hydraulic, and manually operated. They can be clamped with little pressure or tons of pressure, and they can have any shape, dictated by the part. A hex nut can be placed in a hex-shaped hole that has no clamping need but will be held firm because of the part and fixture shape. Fixture design is easy, and only your knowledge of the part and needed processes are required to design fixtures. Many tooling vendors would love to supply you with fixture building materials.

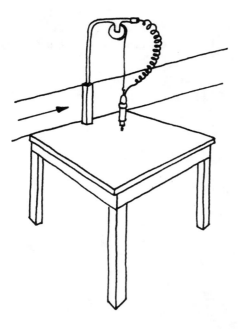

FIGURE 7-6 Counterbalance: Prepositions tool close to the point of use.

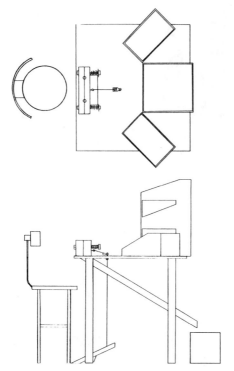

FIGURE 7-7 Fixtures and jigs relieve the hands for more useful work.

FIGURE 7-8 Conveyors move material into and out of work stations.

Use Gravity

Gravity is free power. Use it! Gravity can move parts closer. By putting an incline in the bottom of parts hoppers, parts are moved closer to the front of the hopper. Production management loves us to spare every expense, and the use of gravity can do that. For example, consider a box $24'' \times 12'' \times 6''$ lying flat on a table. The average part in that box (the only part the designer is interested in) as $12''$ back, $6''$ over, and $3''$ down

FIGURE 7-9 Gravity will move parts closer to point of use.

the exact middle of the box. If we get a 2″ × 4″ scrap board out of the trash and place it under the rear end of the box and raise it up 4 to 5 inches, the parts will slide down to the front of the box as the parts are used (see Figure 7-9). The reach has been reduced from 12 inches to 3 inches from the front lip of the box, a significant cost reduction.

Large boxes of parts can be moved into and out of work stations using gravity rollers and skate wheels. Parts can be moved between work stations on gravity slides made of sheet metal, plastic, and even wood.

Gravity can also be used to remove finished parts from the work station. Dropping parts into chutes that carry the parts back, down, and away from the work station can save time and work station space (see Figure 7-7). Slide chutes can carry punch press parts away from the die without operator assistance by using jet blasts or mechanical wipers, or the next part pushes the finished part from the die.

Potential uses of gravity are everywhere. Try to incorporate as much gravity use into your designs as possible.

Operator Considerations

Efficient operators must be allowed to work at the right height, given comfortable chairs, with enough light and adequate space to perform their tasks.

The correct work height is elbow height (see Figures 7-10 through 7.12). With forearm held parallel to the ground and the upper arm straight down, measure the elbow height to the floor. This is the work height. A job should be designed for sitting or standing, but the elbow height must be the same. This requires the designer to calculate

FIGURE 7-10 Wrong work height will create problems.

FIGURE 7-11 Proper work height will produce less fatigue.

working height while standing and then provide a chair that will accommodate that height.

The chair will have to be adjustable. Since work height is dependent on the individual, chairs and tables will have to be adjustable for efficient operation. The chair must also be comfortable (see Figure 7-13). This usually means that it supports the back. Also, a foot ring helps comfort. Comfortable chairs and the option of sitting or standing give the operator a chance to recuperate while working. The end result is more output and less fatigue.

Adequate lighting may not be available in the normal lighting of a manufacturing

FIGURE 7-12 Design work station for sitting and standing, but keep work height constant.

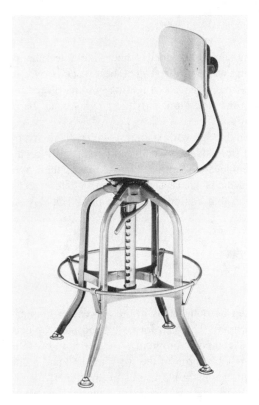

FIGURE 7-13 Chairs must be adjustable and comfortable. (Courtesy of Toledo Furniture).

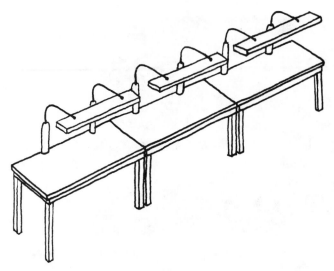

FIGURE 7-14 Adequate, properly positioned lighting will reduce fatigue.

department, so additional lighting should be added—much like a desk lamp (see Figure 7-14). Where to place this lighting is the problem. The best place is over the work and slightly over the back, but not casting a shadow. Lighting is often placed in front of the work, but this causes glare from the reflection. Auxiliary lights could be placed to the left or right of the work.

Operator space is $3' \times 3'$ unless the work station is wider. Three feet off the aisle is adequate for safety, and 3 feet from side to side allows parts to be placed comfortably next to the operator. If two people are working back to back, then 5 feet between stations is OK. If machines need maintenance and clean-up, a 2-foot access should be allowed around the machine. Movable equipment can be placed in this access area if needed for efficient operation.

MOTION PATTERNS

A motion pattern is the path taken by both hands in the process of making one part or pair of parts (if making two at a time). The path for each hand must be unbroken and a complete loop. It is much like a computer program. Another way of defining motion patterns is that a motion pattern is a blueprint of the work method and a bill of material for a time standard. This definition will prove useful in chapter 8, on predetermined time standard systems (PTSS). Work station designs and motion patterns are required before PTSS can be applied, because the motion pattern (which is drawn on the work

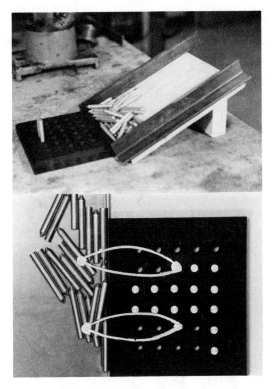

FIGURE 7-15 Work station and motion pattern for pin board: from average pin location to average hole = 6 inches.

station layout drawing) is the blueprint of the work method and the bill of material for the time standard.

The work station design must be completed before the motion pattern is drawn. The first motion pattern is drawn on the work station, and then redrawn on its own page to allow analysis. The motion pattern and cyclograph for the same job would look identical. Cyclographs are nice but too expensive to produce (see Figure 2-2).

Each job is made up of elements. An element is one indivisible piece of work which usually includes a reach, a grasp, a move, and an alignment/position. More simply put, an element is getting one part or tool and doing something with it.

Examples of motion patterns are shown in Figures 7-15 and 7-16.

Notice in Figure 7-15 that there is only one loop per hand. This is a one-element job. In Figure 7-16, there are three loops per hand; therefore, this is a three-element job. Elements are important in setting time standards.

Work station designs using the principles of motion economy and the resulting motion pattern are required before a predetermined time standard can be set. The back of the PTSS form has a place for these two pieces of information. Chapter 8 is dedicated to setting the time standard on the work stations designed in this chapter. Once time standards are set, costs are calculated and all future improvements to the job are compared to these costs.

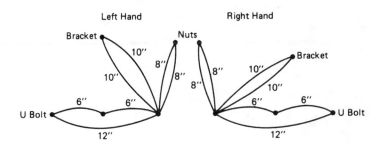

FIGURE 7-16 Motion pattern for cable clamp.

QUESTIONS

1. What is included in a work station design?
2. What is the basic rule regarding cost of new work stations?
3. Where do you start a work station design?
4. What are the principles of motion economy that relate to

      a. Hand motions?

      b. Basic motion types?

      c. Location of parts and tools?

      d. Relieving the hands of work?

      e. Use of gravity?

      f. Operator considerations?

5. What is a motion pattern?

6. What is the use of a motion pattern?

7. Design a work station for your semester project.

8. Draw the motion pattern for your project.

9. Write about how you used the principles of motion economy in your project.

CHAPTER 8

Predetermined Time Standard System

INTRODUCTION

Station design and motion pattern development are the first two steps of the predetermined time standards system (PTSS). Once these two steps are complete, a time standard is needed for the many reasons discussed in chapter 3, but costing the method is most important at this early stage of design. In chapter 7 we asked where to start, and the answer was anywhere, because the first design is never the best design; it is only a starting point for optimization. All future designs will be compared to our first design, and cost is the measure of desirability.

PTSS is a modern technique of motion study and time standards development. All work has been reduced to basic motions, and each motion has been reduced to a specific time value. With PTSS, we can describe the method in terms of the basic motions required to perform the operation, and then the time value for each motion is looked up on a table and recorded on the PTSS form. Each element of work is totaled, and the elemental times are totaled for the time required of the total job.

PTSS is best used before production starts. Think about it: A new product has been invented, and you have been asked to design a production line to produce 2,000 units per shift. There are no machines, no assembly stations, no operators to study. How are you going to set time standards so you will know how many machines to buy or to build, how many people are going to be needed, or how much your product will cost? The PTSS technique is the answer. All new work stations start with an idea in a technician's head. The PTSS will allow the technician to evaluate and improve methods ideas.

PTSS is the efficiency expert's most efficient tool. It can:

1. Develop the best method;
2. Establish the time standard;

3. Help develop standard data (see chapter 10);

4. Establish labor cost;

5. Help justify better tooling;

6. Help select the best machine for the job;

7. Assist in the training of the operator in the best method;

8. Develop motion consciousness and cost consciousness in supervisors, engineers, and employees; and

9. Settle grievances in connection with time standards.

The motion pattern is our blueprint for PTSS. Each line on a motion pattern is either a reach or a move. Larger stations could use body motion, but this is not good station design. (We discuss body motions later in this chapter.) At the end of reaches, we have grasps. Grasps are shown as big dots on the motion pattern. At the end of a move are alignments/positions, releases, or sometimes other grasps. These are also shown as big dots on the motion pattern. A well-defined motion pattern makes PTSS easy. Time spent on motion pattern development will save twice as much in PTSS analysis and development.

The motion pattern is our bill of materials for a time standard. The finished motion pattern is made up of dimensional lines representing reaches and moves, plus big dots representing alignments/positions and grasps. These lines and dots are the sequence of motions required to do a job, and when we assign time to each motion, we end up with a time standard.

We discuss PTSS as follows:

1. Review of the table and definition of terms

2. The form

3. A thirteen-step procedure

4. Example problems.

PTSS TABLE

The PTSS table (see Figure 8-1) is the source of time standards for all motions. The left-hand side of the table is the time standards in thousandths of a minute (.001). The decimal has been omitted; however, if we are consistent and do all work in thousandths of a minute, we will have no problem, at the appropriate time, putting the decimal back where it belongs.

On the right side is the motion pattern construction table (which tells you what motions you can't do at the same time). On the bottom right side are definitions of the terms. Our discussion starts with the definition of each term and a discussion of what causes the time to vary. The table needs to be kept at hand when studying these terms.

FRED MEYERS & ASSOCIATES

PREDETERMINED TIME STANDARDS
NORMAL TIME IN .001 MINUTES

REACHES & MOVES   R or M

2"/.001 + .003 MAXIMUN 48"
+25% FOR EACH 10 # OVER 5

GRASPS - G

G CONTACT GRASP	1
G1 LARGE PARTS 1" OR OVER	3
G2 MEDIUM PARTS 1/4-1"	6
G3 SMALL PARTS UNDER 1/4"	9
G4 REGRASP	4
RL RELEASE	-

POSITIONS & ALIGNMENTS   AP

AP1 LESS THAN 1/4"	5
AP2 LESS THAN 1/32"	10
AP3 LESS THAN 1/64"	20

BODY MOTIONS

SF STATIC FORCE	5
EF EYE FIXATION	5
ET EYE TRAVEL	10
FM FOOT MOTION	5
LM LEG MOTION SAME AS REACH	
B BEND AT KNEES	15
AB ARISE FROM BEND	15
S STOOP TO FLOOR	30
AS ARISE FROM STOOP	30
T1 TURN ONE FOOT DISPLACED	10
T2 TURN TWO FEET DISPLACED	20
ST SIT OR STAND	20
W-F WALK FEET	4
W-P WALK PACES	10

GOALS: ELIMINATE
         COMBINE
         CHANGE SEQUENCE
         DOWNGRADE

MOTION PATTERN CONSTRUCTION TABLE

RIGHT HAND

LEFT HAND		AP1 (-10 / +10)	AP2 (-10 / +10)	AP3 (-10 / +10)	G1 (-10 / +10)	G2 (-10 / +10)	G3 (-10 / +10)
	AP1	- / 1	1 / 2	1 / 1	- / -	1 / 2	1 / 2
	AP2	1 / 2	1 / 2	2 / 3	1 / 2	2 / 3	2 / 3
	AP3	1 / 2	2 / 2	3 / 4	1 / 2	2 / 3	2 / 4
	G1	- / 1	1 / 2	1 / 2	1 / 2	1 / 2	1 / 2
	G2	1 / 2	2 / 3	2 / 3	1 / 2	2 / 3	2 / 3
	G3	1 / 2	2 / 3	2 / 4	1 / 2	2 / 3	2 / 3

CODE 1	1st MOTION FULL VALUE / 2nd MOTION 1/2 VALUE
CODE 2	1st MOTION FULL VALUE / 2nd MOTION FULL VALUE
CODE 3	1st MOTION FULL VALUE / PLUS AN R2 or M2 / 2nd MOTION FULL VALUE
CODE 4	1st MOTION FULL VALUE / PLUS AN ET / 2nd MOTION FULL VALUE

THE BASIC MOTION EMPLOYED TO:

REACH: MOVE THE HAND TO A DESTINATION
MOVE: MOVE AN OBJECT TO A DESTINATION
GRASP: SECURE SUFFICIENT CONTROL TO PERFORM THE NEXT MOTION
ALIGNMENT/POSITION: ALIGN, ORIENT & ENGAGE ONE OBJECT WITH ANOTHER
STATIC FORCE: EXERT FORCE WITH NO MOVEMENT
RELEASE: RELINQUISH CONTROL
EYE FIXATION: FOCUS THE EYES & DETERMINE CERTAIN READILY DISTINGUISHABLE CHARACTERISTICS WITHIN A 16" DIAMETER
EYE TRAVEL: SHIFTING THE EYES

FIGURE 8-1 PTSS table: Time for each motion.

Reach: Symbol R

Reach is the basic motion used to move the hand to a location or destination. Reaching for a part or a tool are good reasons for a reach. We may even be carrying a tool (a pencil, for example), but if our basic motivation is to get to another part or tool, the motion is called a reach. Most of the time, the hand will be empty.

The cause for time to vary is logical: The farther you reach, the more time it takes. In the motion picture analysis of work, it was discovered that the hand moves 2 inches in 1/1,000 of a minute. Motion pictures of work were taken at 1,000 frames per minute. When the film is reviewed at one frame per second, the hand can be seen moving in 2-inch increments across the screen. At the beginning and ending of a reach, there is time required to accelerate and decelerate, about .002 minutes each. If you could eliminate either of these accelerations or decelerations, you could save .002 minutes.

The formula for reaching is 2 inches per .001 plus .003 minutes, up to 48 inches. Beyond 48 inches, the body bends at the same time the hand moves, creating a compound motion that saves time but wears out the operator. Reaches over 36 inches are very fatiguing and should be eliminated.

Examples:

	Code	Reach	Time in .001	Explanation
1.	R1	1″	4	$1 \div 2 + .003$ [a]
2.	R15	15″	11	$15 \div 2 + .003$
3.	R36	36″	21	$36 \div 2 + .003$
4.	R50	50″	27	$48 \div 2 + .003$

[a] Never split .001.

The one complexity to add is that when the reach begins or ends in motion, .002 minutes is subtracted for each.

Examples:

	Code	Reach	Time in .001	Explanation
5.	R10m	10″	6	$10 \div 2 + .003 - .002$
6.	mR20	20″	11	$20 \div 2 + .003 - .002$

The lowercase "m" at the end of the reach indicates ending in motion. Example 5 reads reach 10 inches ending in motion, while example 6 reads beginning in motion reach 20 inches. Unless otherwise noted, reaches begin and end at a stop. Most of our reaches (95%) will be like examples 1 through 4. Anytime the hand changes directions over 120 degrees, the hand comes to a stop.

The hand may reach in a curved motion, but all measurements are made as if the motions were flat. This makes measuring the distances at the work station or on the work station drawing much easier. Just measure point to point.

Move: Symbol M

Move is the basic motion used to move an object to a location or destination. Moving a part from the bin to the fixture, or moving a tool from the table to where needed, are good examples of moves.

There are three causes for move time to vary:

1. Distance measured in inches;
2. Beginning or ending in motion; or
3. Weight or force required.

The first two causes for move time variations are exactly the same as those for reaches. In fact, the time is exactly the same for reaches and moves if the item being moved weighs less than 5 pounds.

For those items (tools or parts) that weigh over 5 pounds, we add 25% more time for every 10 pounds over 5 pounds. If both hands are used, the weight is divided by 2.

Pounds	% Additional Time
5–15	25
15–25	50
25–35	75
35–45	100
45–55	125

Examples:

	Symbol	Move in Inches and Pounds	Time in .001	Explanation
1.	M18	18″	12	18/2 + .003
2.	M6m	6″	4	6/2 + .003 − .002
3.	M20-20#	20″-20#	20	(20/2 + .003)150%
4.	M17-50#/2	17″-25#	21	(17/2 + .003)175%
5.	M11-7#	11″-7#	12	(11/2 + .003)125%

Note: When weight is a factor, no adjustments are made for beginning or ending in motion.

Reaches and moves account for 50% of all operator-controlled work. At this point, you can set time standards for 50% of all work. Try these examples:

LH	Time	RH
R16		R12
M17		M15
R15m		R15m
mM21		mM21
M5-50/2		M5-50/2
M50		M50
R36		M24

Grasps: Symbol G

Grasp is the basic motion used to secure sufficient control of an object to perform the next motion. After a reach to a tool or part, we must secure sufficient control to move that part or tool back to the point of use.

There are five types of grasps:

Contact Grasp: Symbol G. A contact grasp requires no closing of the fingers; it is merely touching something at the end of a reach. Touching a piece of paper before moving it off a desk is an example of a contact grasp. A contact grasp is the fastest motion in all PTSS—.001 minutes. When something must be moved without picking it up, a contact grasp is used.

Large Parts Grasp: Symbol G1. This grasp is used when picking up something which measures at least 1 inch at the point of grasp. The time includes time for the fingers to separate one part from other parts, to close the fingers around that part, and to apply sufficient pressure to gain control.

Medium Parts Grasp: Symbol G2. This grasp is used when picking up parts between $\frac{1}{4}$ inch and 1 inch at the point of grasp. The time for a G2 includes the same motions as a G1, but more because smaller parts are more difficult to handle than larger parts.

Small Parts Grasp: Symbol G3. This grasp is used when picking up parts under $\frac{1}{4}$ inch at the point of grasp. The time for a G3 is the largest time for all grasps because of the smallest dimensions. G3s have time to search, separate, close fingers, and apply pressure to pick up a part or tool. (Think about picking up a straight pin.)

G1s, G2s, and G3s are the normal grasps used in work. The grasps all include time for separation of parts. If this separation can be eliminated, two thirds of the time could be eliminated, and then G1 = 1, G2 = 2, and G3 = 3. When one tool or part is all by itself, it can be picked up using less time. This is a fine point, and failure or inability to design a station and methods so the operator doesn't have to separate tools or parts will not adversely affect the operator's performance.

Regrasp: Symbol G4. A regrasp is used in many different situations and is called the *contingency element*. A contingency is defined as an accidental happening; in PTSS, a G4 is some extra time allowed for a contingency.

The first use is as the name implies: Regrasping a part previously under control. For example, a screw driver when turned 180 degrees cannot be turned any more by hand until it is regrasped. A wrench cannot be rotated past 180 degrees until regrasped. A nut cannot be run down past 180 degrees until regrasped. If a nut is screwed down on a bolt three turns, six regrasps would be needed.

A second use of G4 is palming and unpalming. When more than one part is picked up out of a box of parts, the first part must be palmed to free the fingers before the second part is picked up. If four parts are needed, there would be four G1s, G2s, or G3s and three G4s. The last part would remain in the fingers, moved back to the fixture, and used. When the second part is needed, it is unpalmed and put back in the fingertips where it can be used. There will normally be a palming and an unpalming. Tools are also palmed and unpalmed. Think of a pencil that must be used every cycle. It could be kept in the hand, even if the fingers are needed for something else; it is palmed when not needed and unpalmed when needed.

The third use of G4 is the contingency element. If parts are very small (under $\frac{1}{64}''$), very large (over 10 pounds), fragile, slippery, or dangerous, extra time may be needed. Any grasp or alignment/position with one of these conditions requires one G4 for every condition.

Release: Symbol RL

Release is used when control is relinquished. When a tool is set down, when a part has been used, and when a finished product is asided, the ending element is release. Release has no time value. It is used for description only. If any situation arises where the analyst feels time is needed on the release because of safety, quality, or other considerations, add a G4. The key to good time standards is to be realistic and use common sense.

Alignment and Position: Symbol AP

Alignment is the basic motion required to bring an object to a point or to a line. Position is the basic motion required to align and engage one object with another. Alignment and position are the two motions required to do something with the parts in our hands. The time values are very similar; therefore they have been combined in PTSS. The position includes only 1 inch of engagement; therefore, if a 6-inch engagement is needed, an AP would be followed by an M5. There are three APs:

AP1. is an alignment or position with less than $\frac{1}{4}$-inch tolerance. If an alignment can be within $\frac{1}{4}$ inch, an AP1 is used. If the two parts in a position have $\frac{1}{4}$ inch or less of difference between the dimension, an AP1 is used. If more than $\frac{1}{4}$ inch of clearance is available, no AP is needed, and at the end of a move, only a release is needed.

AP2. is the same as AP1, only with closer tolerance, $\frac{1}{32}$-inch tolerance for alignments and positions. If a pin and hole had a $\frac{1}{32}$-inch difference or less, it would be an AP2.

AP3. is the same as AP1 and AP2, but with still closer tolerance, $\frac{1}{63}$ inch or less. There is nothing less than AP3.

The tolerance for positions is at the point of engagement, as shown in Figure 8-3.

FIGURE 8-2

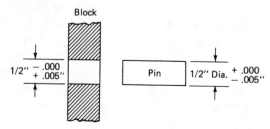

Figure 8-2 is an example of an AP3 and will take .020. The point of engagement is where the top of the block meets the leading edge of the pin when engaged. For cost reduction, the pin will be redesigned to add a champher or a countersink (as in Figure 8-3).

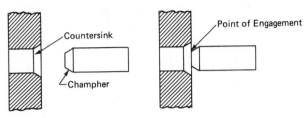

FIGURE 8-3

A $\frac{1}{32}$-inch champher and/or a $\frac{1}{32}$-inch countersink will reduce this to an AP2 and will take out .010 minutes. A 50% savings results. Now, if we champhered and countersunk $\frac{1}{8}$ inch on each, the clearance at the point of engagement will be $\frac{1}{4}$ inch, and an AP1 results. Now the time is reduced to .005 minutes.

APls and AP2s are not too costly, but AP3s are very time consuming. AP3s should be eliminated. Product redesign can eliminate AP3s by downgrading, but tool design and fixturing can totally eliminate the AP1s, AP2s, or AP3s. Let a press put it in. This is especially true of AP3s. Alignments have almost totally been eliminated by fixtures. If a hole is needed in the middle of a steel plate, the tolerances are only .001 inches. The tool designer will provide a fixture with two back pins and one side pin. The operator will move the part into the fixture up against the back two pins and then move it a few inches sideways to the third pin. Now it is located, and no APs were used, just moves. The operator now reaches for the two palm buttons to activate the machine.

Static Force: Symbol SF

A static force is the basic motion used where no movement is involved, but a pressure must be built up. Static force is a tightening or breaking-loose element. If the method requires a tight nut, bolt, or screw, the G4s required to run the part down are followed by an SF. If a bolt has to be loosened, a static force preceded the G4s. Static forces are used to cut wire, press parts together, and loosen faucets (the list will grow with experience). Static force is characterized by a hesitation or lack of motion that takes time. This time must be accounted for with an SF.

Eye Fixation: Symbol EF

An eye fixation is the basic motion used when focusing the eyes on an object to determine certain readily distinguishable characteristics within a 16-inch diameter. The time for eye fixation is normally cancelled by other motions and the time is included only when the hands stop so the eyes can see something that couldn't be seen if the hands and object were moving. Eye fixations are uncommon in production work and common in inspection work. The new technologist should be careful not to use eye fixations when not required. The eyes are always working, and they see most everything without stopping the hands. The best test for when to use an eye fixation is, "Does the hand have to stop for the operator to see this?" The eye fixation is a red flag to a chief industrial engineer, so the technologist must be prepared to state why the eye fixation was used.

Eye Travel: Symbol ET

Eye travel is the basic motion required to shift the eyes from one position to another. Eye travel is required when more than one eye fixation is required and the distance must be over 16 inches apart. The same cautions discussed for eye fixations are true with eye travel—they should only be used when everything else stops. Eye travel is a part of the motion pattern construction table and is discussed further later in this chapter.

Eye fixations and eye travel are the inspection elements. If the quality control procedure calls for checking this and that, the technician counts up the number of things to be checked and allows for that many EFs plus the travel between them if over 16 inches apart.

Body Motions

There are twelve body motions. Our attitude toward body motions is negative, and such motions should be eliminated. Many existing jobs include body motions; therefore, the present method will require the setting of standards using these body motions. The proposed method would eliminate these motions and save a lot of money. The body motions are used in pairs and are discussed that way here.

Foot Motion (FM) and Leg Motion (LM). The foot motion is the basic motion used to pivot the foot at the heel, while the leg motion is the basic motion used to pivot the leg at the knee or hip. Both motions are measured in inches and are calculated just like reaches and moves. Foot motion is a useful and efficient body motion as long as the distances are minimized. Leg motions should be eliminated or at least reduced to foot motions. Microswitches can be activated with a 1-inch foot motion and are easily cost justified. Pressing the gas pedal on the car is a foot motion, while moving to the brake pedal is a leg motion. Some people ride with their foot over the brake at all times. Do you think they are industrial technologists?

Bends (B) and Arise from Bend (AB). A bend involves reaching down (to about knee level) to pick up a part or tool. Once at knee level, the operator will pick up or put down a part or tool and then arise from bend. Bending and arising from bends move the hands about 24 inches, which takes .015 minutes; therefore bends and arise from bends comprise .015 minutes. Bending the body down to around the knees several times per minute will quickly fatigue the worker. For this reason, bends and arising from bends should be eliminated. Parts that come to a work station in $4' \times 4'$ shop tubs will require a bending motion to pick up the average part (the one in the exact middle of the tub). A methods improvement would bring the parts to the work station on a conveyer at waist level. This would eliminate the bending and the arise from bending, a savings of .030 minutes.

The position of the average part is always used because only one cost or standard per operation is allowed. The top parts on a tub will be faster, and the bottom parts will be slower, but the one in the middle is the average time. PTSS is better than time study because of this fact.

Stoop (S) and Arise from Stoop (AS).

Stooping is the basic motion used to reach down around the ankles, and arise from stoop is coming back upright. When parts or tools must be picked up from the floor, stoop and arise from stoop are required. Stooping is extremely fatiguing and must be eliminated. Bringing parts to the work station on elevated material handling devices is the answer. The cost of the material handling device will be paid for by the reduction in time resulting from the improvement. Stoop is the maximum reach (48 inches) plus a 10% fatigue allowance.

Turning: Symbol T1 or T2.

Turning is the basic motion of displacing one foot up to 110 degrees. If an operator must turn sideways, one foot is displaced, and this is called a T1 (one foot displaced). If the operator must turn around, two feet must be displaced, and this is called a T2. Work stations that require operators to turn are not properly designed, and opportunities to reduce costs are available. A T2 moves the part 34 inches, and the time is .020 minutes. A T2 also takes two paces, and that time is .020 minutes.

Sit and Stand: Symbol ST.

Sitting and standing motions are considered from a shop chair (about 36-inch seat height). No work station designer would design a job where the person had to stand up each time and sit down again during the same cycle, but someone may have in the past. That is the only reason for knowing how much time sitting and standing should take. All work stations should allow the employee to work sitting and standing alternately to reduce fatigue, but this does not require the operator to get down and up in the same cycle. Normally, no time is allowed for the operator to change positions for comfort. This time is in the allowance addition (discussed in chapter 9).

Walking: Symbol W__F, W__P.

Walking can be measured in feet (symbol F) or paces (symbol P); therefore, W10F is read, walk 10 feet, and W5P is read, walk 5 paces. W10F equals $10 \times .004$ minutes or .040 minutes, and W5P equals 5×10 or .050 minutes. Walking time standards are based on the universally accepted standard of 264 feet/minute, or 3 miles per hour, and rounded off since we don't split .001. $(\frac{1}{264} = .00378$ minutes.) A pace is 30 inches.

MOTION PATTERN CONSTRUCTION TABLE

The top right-hand side of the PTSS table is called the motion pattern construction table. It tells us what we can do at the same time and what we can't do. Reaches and moves can be done two at a time for just the time cost of one reach or move. Reaches and moves are not on the motion pattern construction table; those symbols not included on the table can be done two at a time. Only alignments, positions, and grasps are included. Anytime the work method asks the operator to perform two alignments, positions, or grasps at the same time, we must look to the motion pattern construction table for guidance.

The motion pattern construction table is arranged like a mileage map. The intersection of two motions is a box with two numbers in it. These numbers are code numbers. The top number is used when the hands are working within 10 inches of each other (for example, picking up a part with each hand in the same container). The bottom number is used when the hands are working over 10 inches apart. When the hands are over 10 inches apart, more time is required, and the motion pattern construction table tells us how much more time.

Until this point, all the numbers on the PTSS table have been in .001 minutes. The motion pattern construction table is in code. Four codes are used.

Code 1: First Motion Full Value, Second Motion Half Value

If the motions are not on the chart, or there is no number in the box, both motions can be done at the same time; and on the form, both motions would be on the same line. A Code 1 would be on two lines. For example, grasp two large parts, hands under 10 inches apart.

	LH	*Time*	*RH*	
Grasp part	G1	3		
		2	G1	Grasp part

The time does not go on the same line, because it does not happen at the same time. Also, note that half of three is two, because we don't split .001 minutes.

Code 2: First Motion Full Value, Plus Second Motion Full Value

This code requires each hand to work separately. For example, place two parts in holes over 10 inches apart.

LH	*Time*	*RH*
AP2	10	
	10	AP2

Note that either hand can go first; the motion pattern would tell you what each hand is doing, and common sense is the only guide to which works first. Normally, it doesn't make any difference.

Code 3: First Motion Full Value, Plus a 1-inch Reach, or a Move 1 Inch, Plus Second Motion Full Value

After we have grasped a small part in one hand, our other hand must finish its reach to the other part; therefore, it would be a 1-inch reach followed by the second grasp. For example, grasp two small parts over 10 inches apart.

LH	Time	RH
G3	9	
	4	R1
	9	G3

As another example, if two parts are being assembled, then the second motion would be preceded by an M1.

LH	Time	RH
AP3	20	
	4	M1
	20	AP3

The R1 or M1 is just finishing the reach or move. When the next motion is a grasp, an R1 is used. When it is aligning or positioning, the M1 is used. Both motions have the same time value (.004).

Code 4: First Motion Full Value, Plus an Eye Travel, Plus Second Motion Full Value

When two complicated motions are performed at the same time and the first motion is completed, the eyes must move before that second motion can be performed. Code 4s are always over 10 inches apart; therefore, they are the bottom number in the box. For example, grasp a small part and position a tight fit over 10 inches apart.

LH	Time	RH
G3	9	
ET	10	ET
	20	AP3

Notice that the hands do not have to be doing the same thing, but they cannot do a G3 or an AP3 at the same time.

Now that the definitions have been given, as well as the instructions for their use, practice is needed. The PTSS form has been designed to help the technologist develop the method and standard. The next step in understanding PTSS is to understand the form.

THE FORM

Figure 8-4 shows a PTSS form with circled numbers in each block. In this section, we describe what goes in each block. In an earlier section (see p. 64), we discussed in

FIGURE 8-4 PTSS form: The step-by-step form

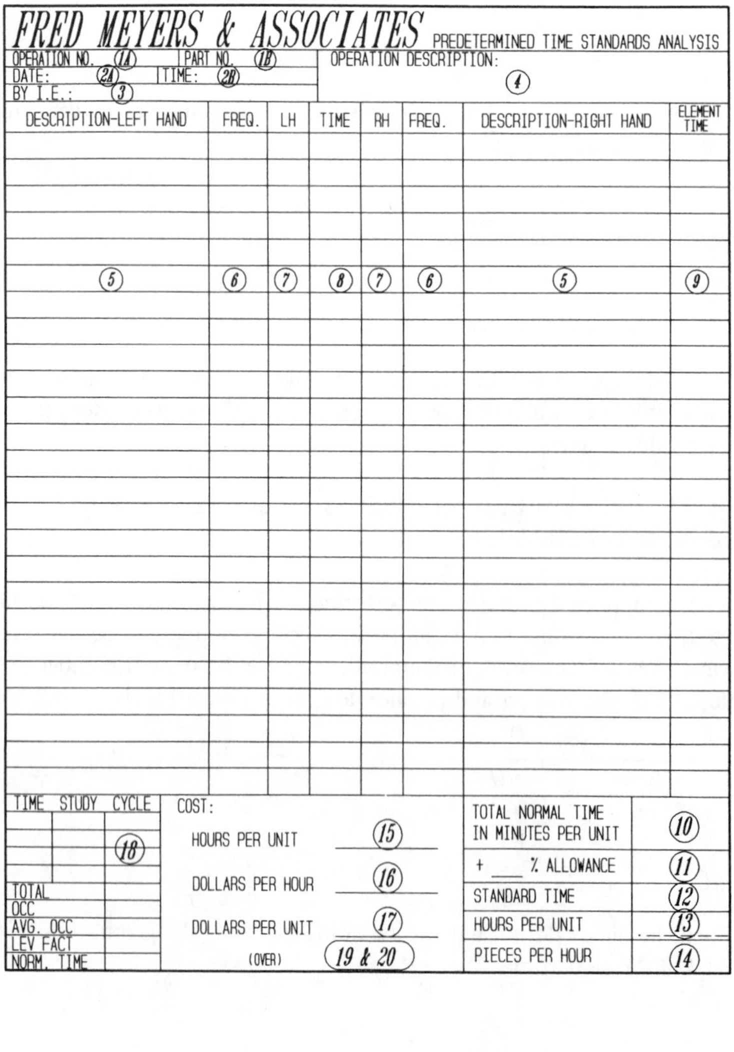

detail much of the information asked for in the top block: part number, operation number, etc. That discussion should be reviewed, because it will not be repeated here.

①.A *Operation Number*   ①.B *Part Number*

Both the operation number and part number are needed to identify the specific job being worked on. If either number is omitted, the study will be valueless.

②.A *Date*   ②.B *Time*

The date, including the year, is important to keep track of the sequence of methods. PTSS studies tend to stay around for years, and it is important for further study to know how old a study is. The time of day is important for the time study because it helps explain the leveling factor, such as late in the day may explain a low leveling factor because of fatigue.

③. *By I.E.*

The industrial technologist will commonly be given the title Industrial Engineer by his or her company. The industrial technologist's name goes in this block.

④. *Operation Description*

This is a brief but accurate description of the work being performed at this station. Key words like *assemble, weld, mill, packout,* etc. are important.

⑤. *Description—Left Hand or Right Hand*

This is where the best description of the operation goes. The elements have been reduced to the simplest motions, and a description of that small piece of work should be easy. This description is important for understanding.

⑥. *Frequency*

Frequency is on both the left-hand and right-hand sides of the form. This column is extremely useful and will save you time. Frequency asks how many times you want to perform this motion. For example, if a nut is required to be turned down eight revolutions, sixteen G4s would be needed. Instead of writing 16-G4 one after the other, 16 is placed in the frequency column and multiplied by the time value for one G4. The total (64) is placed in the time column. At other times, an element may be repeated. Instead of rewriting the total element, the technologist only has to place a 2 in the RH frequency column next to the total element time, then multiply the total element time by the 2, and enter the total in the element time (9) column. An inspection element may not be repeated every cycle. In this case, we calculate the total time required to inspect one part and divide that time by the frequency of inspection called for by the quality control department. 1/10 means to inspect one out of every 10 parts; therefore, 1/10 of the time would be included in the time standard. This philosophy is used for material handling, clean-up, loading parts, etc.

⑦. *LH, RH*

The left-hand, right-hand columns are for symbols. The symbols are meaningful and reduce the number of words needed in the description column. An M16-50/2 means move 16 inches with a 50-pound object using two hands; therefore, this doesn't need to be a part of the description. What is needed in the description would be the part name being moved and the destination. A G3 means pick up a small part, but the part name is missing.

⑧. *Time*

There is only one column for time because many motions can be done two at a time (one in each hand). If the motions are too complicated to do two at a time, two lines must be used. The motion pattern construction table will be our guide to what is too complicated to do at the same time. The time is in .001 of a minute, but the decimal is omitted in this column for simplicity. Every line is a moment in time, and if two motions cannot be performed at the same time, they cannot be on the same line. The motion times are totaled for every element under the last motion. A release is a good ending point.

Example:

Description—Left Hand	Freq.	LH	Time	RH	Freq.	Description— Right Hand	Element Time
Get Braces and Pack Part #1 and 2							
To Part #1		R48	27	R48		To Part #2	
Grasp 2	2	G2	12				
			8	R1	2		
			12	G2	2	Grasp 2	
Regrasp 1st		G4	4	G4		Regrasp 1st	
To Box		M24	15	M24		To Box	
In Box		RL	—	RL		In Box	
			78				.078

⑨. *Element Time*

The element time is the total of all the motions of one element of the job. The number of elements is equal to the number of legs on a motion pattern. An element usually includes a reach, grasp, move, alignment/grasp, and a release for each hand. The motion times for each of these were totaled in the time column. Our previous example showed doing this element twice, so two times the element is placed in the element time column. Up to this point, no decimal was needed. Now the decimal is placed in the third place from the right—1.307, for example.

⑩. *Total Normal Time*

The total normal time is simply the total of the element times. If there are three elements, there will be three element times. These elements are added together for the total normal time. Normal time is defined as the amount of time required for a person to perform a specific task at a normal pace. There are no allowances in normal time.

⑪. +__% *Allowance*

Plus 10% allowance will be used until our discussion in chapter 9. Allowances are that extra time allowed in each cycle for getting tired, personal needs, and unavoidable delays. Ten percent of the previous number (total normal time) is placed here.

⑫. *Standard Time*

Standard time is normal time ⑩ plus allowances ⑪.

⑬. *Hours/Unit*

Hours per unit equals standard time ⑫ divided by 60 minutes per hour.

⑭. *Pieces/Hour*

Pieces per hour equals the whole number 1 divided by hours/unit ⑬. ⑬ and ⑭ are $1/x$ of each other. This is called the reciprocal.

Examples:

⑩ Normal Time	+	⑪ 10% Allowance	=	⑫ Standard Time	⑬ Hour/Unit	⑭ Pieces/Hour
.200		.020		.22	.00367	273
.520		.052		.572	.00953	105
.725		.073		.798	.01330	75
1.000		.100		1.100	.01833	55

⑮. *Hours/Unit*

Same as ⑬. Now being used for costing.

⑯. *Dollars per Hour*

The wage rate of the operator. Most often an average wage rate of a department. The loaded labor rate means that the labor rate has fringe benefits included.

⑰. *Dollars per Unit*

The dollars per unit is the labor cost of one unit of production using the method described. Dollars per unit is used to compare methods.

⑱. *Time Study Cycle*

The time study cycle is a small time study form. Room for twelve cycles has been made. (Time study is discussed in chapter 9.) What we want here is normal time to compare to ⑩, total normal time. An industrial technologist can feel comfortable if the time study normal time agrees with the PTSS normal time. Often, however, the work station is not available for time study until months after the PTSS standard has been set. However, when it is available, a time study should be made.

⑲. *Layout*

The time standard cannot be set without a layout. Layout of the work station comes first, and a large area on the back of the PTSS form has been assigned for this purpose. (In chapter 7, we discussed work station design.)

⑳. *Motion Pattern*

The motion pattern is the second step of PTSS. After the work station layout is completed, a motion pattern is drawn. The motion pattern is the blueprint for the PTSS method and the bill of materials for the time standard.

This twenty-item list is a reference for what goes in each block. Refer to it when you are not sure what is needed.

The step-by-step procedure discussed next will assist you in understanding the PTSS form.

THIRTEEN-STEP PROCEDURE FOR DEVELOPING A PTSS STUDY

Step 1: Select an operation to study

A new product, a request from management, or an industrial technologist who has a cost reduction idea are all possible sources of new projects to study. There are always opportunities to study work.

Step 2: Collect data

The data needed to design a work station includes blueprints, bill of materials, and production volume. The technologist needs to understand what work must be accomplished. During the process design stage of a new product, route sheets must be developed for each part, and the sequence of assembly and packout is developed. Each of these operations (as shown on an operations chart) requires a work station design. The more information the designer has, the better job he or she can do.

Step 3: Lay out Work station

We discussed work station layout in chapter 7. Nothing can be done until this step is complete. If the work station layout changes, a new time standard will result.

Step 4: Develop a Motion Pattern

We discussed motion pattern design in chapter 7. The motion pattern is shown on the station layout, and the motion pattern becomes the blueprint for the motion study analysis.

Step 5: Break Down the Job into its Smallest Motions

The motion pattern developed in Step 4 is the blueprint. (See Figure 8-5 and 8-6.) The motion pattern shown in Figure 8-5 is for assembling two "U" bolts. The motion pattern has broken down the job into three elements—three loops. Note that the cyclograph (shown in Figure 8-6) looks just like a motion pattern. This cyclograph is for a five-element job.

Step 6: Break Down the Job into Elements

The motion-pattern shown in Figure 8-5 has three elements. Note the three loops for each hand. Each loop is a part. Since both hands are working at the same time, this is a three-element job. An element of work cannot be subdivided between operators. If this were part of a larger assembly, one part (or element) could be taken from one person and given to another as the volume of output needs change. Each element will be subtotaled on the PTSS form

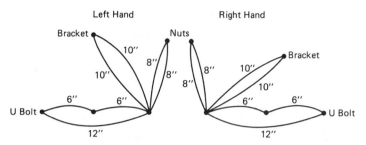

FIGURE 8-5 Motion pattern: Three-element job.

and carried over to the elemental time column with the properly placed decimal—three places.

Step 7: Calculate Allowances

Allowances are extra time added to the time standard to allow for coffee breaks, personal time, and unavoidable delays. A typical allowance is 10% in addition to the work content time. At this point, a constant 10% is added to the normal time for allowances. In chapter 9, we discuss allowances in more detail.

FIGURE 8-6 Cyclograph: Five-element job.

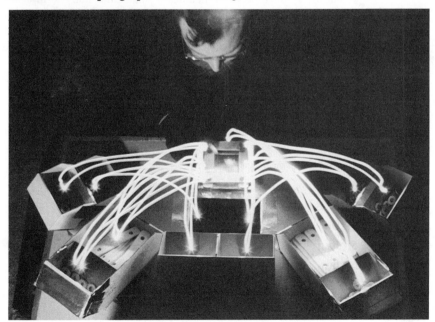

Step 8: Calculate the Time Standard

A time standard consists of three numbers:

1. *Standard time:* Normal time plus allowances equals standard time in minutes.

2. *Hours per unit:* Hours per unit is the standard time (in minutes) divided by 60 minutes/hour. This standard is always communicated in five decimal places (.00001 hours). Because this number is not meaningful to most people, hours per 1,000 units is often used. In that case, two decimal places are used—0.01 hours per 1,000.

3. *Pieces per hour:* Hours per unit are divided into the whole number 1 to calculate pieces per hour. This is the number most people think of when talking about time standards.

Step 9: Calculate Cost:

The hours per unit multiplied by the hourly rate of the operator equals the unit labor cost. This labor cost is used to compare all future methods improvements.

Step 10: Time Study Comparison

When a work station exists, a stopwatch time study can be done and used to compare the PTSS time standard. Time study is the subject of chapter 9, but the PTSS time form includes a small time study form at the lower left. When the PTSS normal time agrees with the stopwatch normal time, the technologist can be confident that the time standard is correct. If they don't agree, the technologist will look for problems that were overlooked in either time study or PTSS.

Step 11: Improve for Cost Reduction

Once the first ten steps are complete, we have a method and a cost. There is always a better way, so the technologist will search for a better method. The technologist will try to

1. Eliminate motions.

2. Combine motions together.

3. Change the sequence of motions.

4. Downgrade the motions to less time-consuming ones.

5. Justify new equipment.

6. Justify better tools, fixtures, and jigs.

Step 12: Select the Best Method

The best method is the cheapest unit cost method considering all costs. Expensive and fast equipment can be justified only if enough product is needed.

High-volume production can support high-cost tooling, but low-volume production can only afford minimal tooling and machining expense.

The cost of the proposed method is subtracted from the cost of the present method, and the result is multiplied by the planned yearly production. This is the annual savings. The cost of the new proposal is then divided by the annual savings, resulting in a return on investment (ROI). An acceptable ROI varies from 25% (four-year payback) to 100% (one-year payback), depending on company goals and policy. A 100% ROI is an almost certain approval.

Step 13: Publish the Approved Method and Time Standard

The company will have a system for communicating the method and standard to all who need them. The approved method and standard will be used for all uses of time standards discussed in chapter 1.

EXAMPLE PROBLEMS (FIGURES 8-7 THROUGH 8-10)

Operation 2010 on the swing set packout line requires the operator to pack out three different parts listed on the back of the example. The work station layout shows the operator bending down into tubs to retrieve parts. The motion pattern is for both hands.

This is a poor method, and an improved method follows (see Figure 8-9).

Study these two examples closely. This plant needs to produce 300,000 swing sets per year. How much will the proposed change save the company?

QUESTIONS

1. What will PTSS do for you?
2. What are the definitions for each motion?
3. What causes time to vary for reaches, moves, grasps, and alignment/positions?
4. What is the motion pattern construction table?
5. How do we eliminate AP3?
6. What is our attitude toward body motions?
7. How do we use the frequency column on the PTSS form?
8. What are the decimal rules?
9. Why do we divide the job into elements, and what is an element?
10. What are the thirteen steps of the PTSS procedure?
11. Complete a PTSS study for your semester project. At least four elements and .250 minutes are required. Subtotal each element.
12. What is the annual savings for Figure 8-7 with thirty assemblers and Figure 8-9 with fifteen assemblers?

FRED MEYERS & ASSOCIATES PREDETERMINED TIME STANDARDS ANALYSIS

OPERATION NO. 2010	PART NO.				OPERATION DESCRIPTION:	
DATE: 10/19/xx	TIME:				Packout Parts #1. 3. & 6.	
BY I.E.: Meyers						

DESCRIPTION-LEFT HAND	FREQ.	LH	TIME	RH	FREQ.	DESCRIPTION-RIGHT HAND	ELEMENT TIME
Packout Part #1--2each							
			10	T1		Turn Left	
			40	W10F		Walk 10 feet	
			15	B		Bend	
			6	G2		Grasp #1	
Grasp #1		G2	6				
			15	AB		Arise	
Move to Box		M36	21	M36		Move to Box	
in Box		RL	--	RL		in Box	
			113				.113
Packout Parts #3 & 6 = 2 #3 & 1 #6							
			20	T2		Turn Around	
			15	B		Into Tub #3	
			3	G1		Grasp #3	
Grasp (1) #3		G1	2				
			15	AB		Arise	
			20	W2P		Side Step Twice	
			15	B		Into Tub #6	
			3	G1		Grasp Part	
			15	AB		Arise	
			20	T2		Turn Around	
		M12	9	M12		Move Parts to Box	
		RL	----	RL		Into Box	
			137				.137

TIME STUDY CYCLE			COST:		TOTAL NORMAL TIME IN MINUTES PER UNIT	250
.30	1.55	2.70				
.58	1.80	3.00	HOURS PER UNIT	.00458		
.91	2.10				+ 10 % ALLOWANCE	.025
1.21	2.40		DOLLARS PER HOUR	10		
TOTAL		3.00			STANDARD TIME	.275
OCC		10				
AVG. OCC		.300	DOLLARS PER UNIT	.0458	HOURS PER UNIT	0 0 4 5 8
LEV FACT		85	# People on line	30		
NORM. TIME		.255		$1.374	PIECES PER HOUR	218

FIGURE 8-7 PTSS: Present method—three tubs of parts: Methods and time analysis.

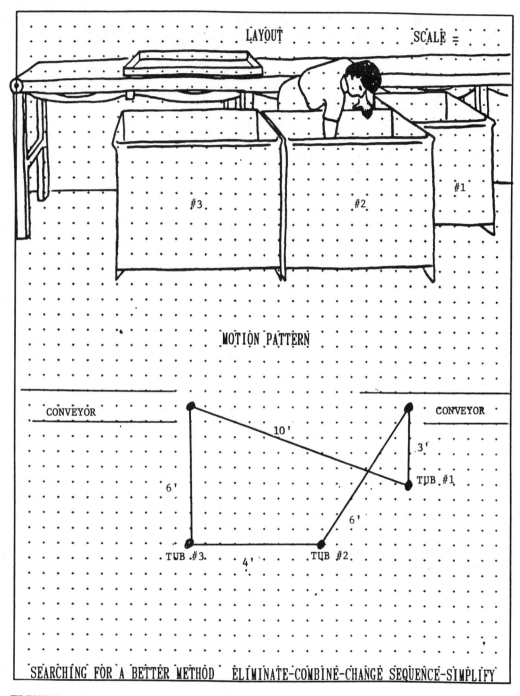

FIGURE 8-8 PTSS back page: Present method—work station layout and motion pattern.

FRED MEYERS & ASSOCIATES PREDETERMINED TIME STANDARDS ANALYSIS

OPERATION NO. 2010	PART NO.		OPERATION DESCRIPTION:
DATE: 10/19/xx	TIME:		Packout 2 each #1, 2, 3, 4, and 1 each of 5 & 6.
BY I.E.: Meyers			

DESCRIPTION-LEFT HAND	FREQ.	LH	TIME	RH	FREQ.	DESCRIPTION-RIGHT HAND	ELEMENT TIME
Get Braces and Pack Part #1 and 2							
To part #1		R48	27	R48		To Part #2	
Grasp 2	2	G2	12				
			8	R1	2		
			12	G2	2	Grasp 2	
Regrasp 1st		G4	4	G4		Regrasp 1st	
To box		M24	15	M24		To box	
In box		RL	---	RL		In box	
			78				.078
Packout part #3 and 4							
To bag		R12	9	R12		To cap	
Grasp 2 bags	2	G1	6		2		
			6	G1		Grasp 2 caps	
Regrasp 1st		G4	4	G4		Regrasp 1st	
To box		M12	9	M12		To Box	
In box		RL	--	RL		In box	
			34				.034
Packout part #5 and 6 1 each							
To #5		R24	15	R24		To #6	
		G1	3				
			3	G1		Grasp #6	
To Box		M24	15	M24		To Box	
		RL	36	RL			.036

TIME	STUDY	CYCLE	COST:		TOTAL NORMAL TIME IN MINUTES PER UNIT	.148
.15	.75	1.35				
.30	.90	1.50	HOURS PER UNIT	.00272		
.45	1.05				+ 10 % ALLOWANCE	.015
.60	1.20		DOLLARS PER HOUR $10.00			
TOTAL		1.50			STANDARD TIME	.163
OCC		10			HOURS PER UNIT	0 0 2 7 2
AVG. OCC		.150	DOLLARS PER UNIT $ .0272			
LEV FACT		95%	#People on line = x15		PIECES PER HOUR	368
NORM. TIME		.143	$.408			

FIGURE 8-9 PTSS form: Proposed packout method and time analysis.

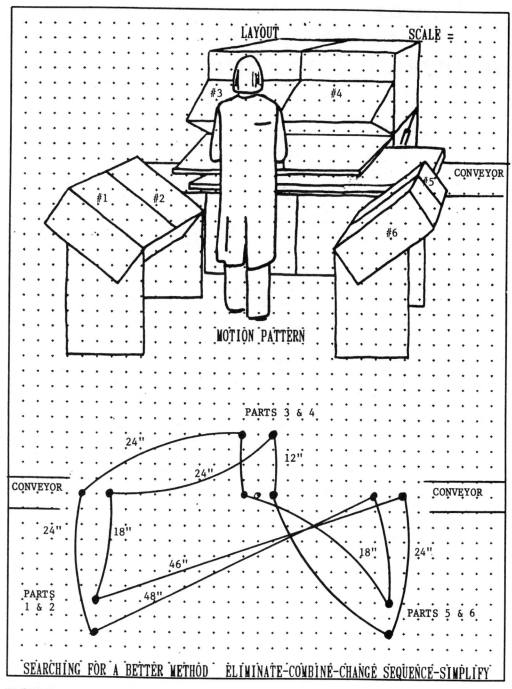

FIGURE 8-10 PTSS form: Back side work station layout and motion
pattern for proposed packout method.

CHAPTER 9

Stopwatch Time Study

INTRODUCTION

Stopwatch time study is the most common technique for setting time standards in the manufacturing area. The time standard is the most important piece of manufacturing information, and stopwatch time study is often the only method acceptable to both management and labor. Stopwatch time study was developed by Frederick W. Taylor in 1880 and was the first technique used to set engineered time standards. Because of this long history, many companies have negotiated stopwatch time study into their labor contracts. Stopwatch time study may not be the best technique for setting a particular time standard, but it may be the agreed-on method to be used. Most unions know management's strong feelings and need for time standards, and they have accepted it as a fact of life. Unions have also developed time study programs and trained people in time study. There is no real conflict between a skilled union time study person and a skilled company time study person. Both are using the same scientific base. Differences do occur, but they are easily resolved between people who know what they are doing and whose goal is to be fair. The 100 plus years of work with stopwatch time study has deeply entrenched it as the technique for setting time standards.

Stopwatch time study is a difficult job only because of some employees' negative attitudes. The time study technician is under pressure from both labor and management. There is humor in the following statement, but a little truth too:

If time study technicians set time standards too tight, labor is mad at them. If time study technicians set time standards too loose, management is mad at them. If time standards are perfect, everyone is mad at them.

114

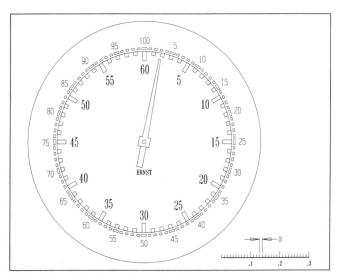

FIGURE 9-1 Stopwatch face: Decimal minutes and seconds

The time study technician's only defense is that he or she knows only one way to set a time standard—the right way. The right way is fair and equitable to all. What everyone really wants is a *good* time standard.

Nevertheless, stopwatch time study is a difficult job. Many graduates of industrial engineering programs think stopwatch time study is beneath them. This is a mistake. Stopwatch time study is the most basic technique of industrial engineering and the best way to learn the company's technology. The industrial technologist is well prepared for this job, and it is a great training ground for other industrial or manufacturing engineering positions leading to upper production management. Six months to two years of experience in time study would make a great foundation for an industrial engineering career. An industrial technologist would find more meaningful work experience applying time standards, as discussed in chapter 2, but knowing how to set time standards will always be a prerequisite to success in all fields of industry.

Figure 9-1 shows a stopwatch. It is in decimal minutes, but unlike PTSS, time is measured in one hundredths (.01) of a minute instead of one thousandths (.001) of a minute. Practice in reading the watch while the watch is running is needed to minimize the industrial technologist's error. A classroom experience can be developed simply by tapping on the blackboard on a predetermined schedule. The student would read the watch and record the reading. Differences should be held to plus or minus .01 minute. If access to an industrial plant is possible, the new technician could time an automatic machine, or semiautomatic machine, such as a punch press. These kinds of operations are consistent, so any fluctuation is the industrial technologist's error. You will be the only one to know how accurate your readings are. To be the best time study technician possible takes practice.

This chapter is organized as follows:

Tools of stopwatch time study

Step-by-step time study procedures

Rating, leveling, or normalizing

Allowances and foreign elements

Time study practices and employee relations

Long cycle time study.

To eliminate redundancy, review the sections "What Is a Time Standard?" in chapter 3 and "Stopwatch Time Study" in chapter 4. These are two important subjects and must be understood before proceeding.

The time study procedure in this chapter starts with a standardized work station design and a skilled, well-trained operator. Both of these conditions have been discussed in previous chapters, and a time study without either is useless.

TOOLS OF STOPWATCH TIME STUDY

The tools of stopwatch time study are important to know before we get into the technique itself, because they play such an important part. The tools discussed in this chapter are as follows:

1. Stopwatches
   a. Continuous
   b. Snapback
   c. Three-watch
   d. Methods time management
   e. Digital
   f. Computer
2. Boards for holding watches and paper
3. Videotape recorders
4. Tachometers
5. Calculators
6. Forms
   a. Continuous
   b. Snapback
   c. Long cycle.

Stopwatches

Continuous Time Study: The continuous time study technique is the preferred technique. Once started, the decimal minute stopwatch will continue to run until the study is completed. The industrial technologist will read and record the element ending time, and the end of one element is the beginning of the next element. The continuous stopwatch has two dials (see Figure 9-2). A large dial is divided into hundredths of a minute (.01) with one revolution of the sweep hand being one minute. The small dial records minutes up to 30. The crown of the watch, when depressed, stops the watch. The crown can be depressed a second time to reset the watch back to start, and a third depression will restart the watch. Three depressions of the crown is time consuming and should be done only once per study.

The continuous time study must be extended when the readings are complete. Since only the ending elements have been recorded, every reading must be subtracted from the previous reading to calculate the elemental time. This subtracting is time consuming and is the most undesirable part of continuous time study.

The continuous time study is the preferred technique of unions, because everything that happens during the study is recorded and open for discussion. This technique is said to have integrity. The time and nature of all foreign elements are recorded, and a decision will be made to include or exclude these elements. (A later section of this chapter covers foreign elements.)

FIGURE 9-2 Continuous stopwatch. (Courtesy of Meylan Corporation)

FIGURE 9-3 Snapback stopwatch. (Courtesy of Lafayette Instrument Company)

Snapback Time Study: The snapback time study technique is faster and easier than continuous time study. Each time an element ends, the technologist reads the watch and immediately snaps it back to zero, and the watch restarts automatically timing the next element (see Figure 9-3). The technologist records the reading at his or her leisure during the next element. The advantage of the snapback is that each recorded reading is the elemental time, and no subtraction is needed. This is a big advantage to the industrial technologist, but the fact that there is no way of checking the industrial technologist's work makes it unacceptable to most unions. It is said that the snapback stopwatch has no integrity.

The crown of the snapback watch resets the watch every time it is pushed. The snapback watch also has a side shifter lever which turns the watch off and on. This is a useful device for when interruptions occur; the time study technician can turn off the watch until the operator goes back to work. But again, there is the chance of error or controversy. The snapback technique is still a useful way of checking time studies and for quick studies.

Three-watch Time Study: The three-watch time study technique is the best of both the continuous and the snapback techniques. Three continuous stopwatches are used on one board. Remember that the first time you push the crown stops the watch, the second time resets the watch, and the third time restarts the watch. On the 3-watch board, there is a watch in each stage (see Figure 9-4). When the operator finishes an element of work, the time study technician pulls a common lever which depresses all three crowns.

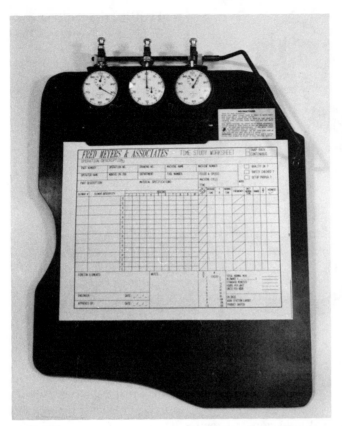

FIGURE 9-4 Three-watch time study method. (Courtesy of Meylan Corporation)

One watch is stopped so a reading can be made, the second watch is restarted and is timing the current element, and the third watch is reset to zero and is waiting to time the next element. The technologist is able to read the watch while it is stopped. A higher-quality reading results, but the big advantage of the three-watch time study technique is the lack of subtraction to calculate elemental time.

The disadvantages of the three-watch technique include the cost of the equipment and the lack of an audit trail—the technologist could leave out elements and no one would know.

Methods Time Measurement Time Study: The methods time measurement (MTM) time study technique is used where the predetermined time standard system is MTM. MTM was developed in the 1940s and was one of the first predetermined time standard systems. Today, there are active chapters of the MTM associations in every industrialized nation of the world. MTM is the grandfather system, but the association keeps it modern. MTM is a copyrighted system, and the association sponsors 80-hour courses

FIGURE 9-5 MTM stopwatch: Reads in TMUS. (Courtesy of Meylan Corporation)

in several of its techniques. It awards a blue card on successful completion of its programs, and many companies require that their industrial technologists have a blue card. The only way to get a blue card is to go to the association's schools. Big companies have in-plant MTM instructors and courses.

The MTM stopwatch (see Figure 9-5) measures time in (.00001) one hundred-thousandths of an hour (.00001), or one TMU (time-measured unit). The hand of the watch makes one revolution in .001 hours (3.6 seconds). This is the same unit of time as in the MTM system, and it is used to reduce the math required. It is efficient, but few people understand what is happening, therefore creating more distrust of industrial engineering and technology.

The sweep hand of the decimal hour stopwatch moves much faster than a decimal minute stopwatch. Upon first observing the MTM watch running, the technologist may think the watch is broken.

Digital Stopwatches: New technology is improving digital stopwatches every year. The big advantage of digital is improved accuracy. The digital watches read in one thousandths (.001) of a minute, and when a read button is pushed and held, an elemental time is displayed (see Figure 9-6). When the read button is released, the memory catches up with the present elemental time and can be depressed again immediately to record another element. Some digital stopwatches have two displays, one continuous and one snapback. These watches are the best of both worlds and have enjoyed increased use every year.

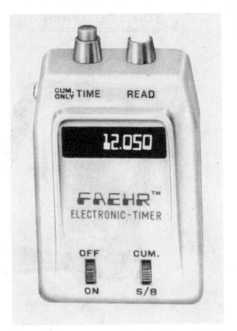

FIGURE 9-6 Digital timer. (Courtesy of Faehr Electronic Timer Company)

Figures 9-7 & 9-8 are examples of digital watches built into time study boards. The Sunlite (L.C.D. display) can be used in bright light.

Digital watches are electronic and need to be recharged. The maintenance of the charge is important.

Digital watches can be used in both continuous and snapback time studies.

Computers: Computers can be programmed to do time study. Several commercial microcomputer time study programs and specially built hardware are available. The advantages are that they are fast, accurate, and can reduce time study cost.

Hand-held data collectors are taken to the shop floor where the time study is performed (see Figure 9-9). The Datamyte 1000 data collector is connected to a computer terminal for extending the time study (see Figure 9-10). The math is totally automatic. Once ending elements are defined and the time study started, the time study technician needs only push a button on the collector to record the time. Accuracy is much improved.

Of all the stopwatches, the continuous watch is used most. Digital watches can replace the older sweep hand, but it is still the continuous time study technique. Computers will soon take over.

Boards

Time study boards vary from cheap clipboards to multiwatch digital boards, and they have one goal—to hold equipment for ease of use.

INDUSTRIAL TIMER

Battery Indicator

Controls
out of way
on side of
case.

Total Time

Event Time

L.E.D. displays

SUNLITE TIMER

Battery Indicator

Controls
out of way
on side of
case.

Total Time

Event Time

L.C.D. displays

FIGURE 9-7 and 9-8 Digital stopwatches. (Courtesy of Faehr Electronic Timer Company)

FIGURE 9-9 Computer time study: Hand-held data collector taken to the shop floor. (Courtesy of Datamyte)

FIGURE 9-10 Computer/VCR time study: Perfect record of what was timed.

If a simple clipboard is used, the watch must be held in the same hand. This is not impossible, but it is not comfortable (see Figure 9-11).

Continuous and snapback time study boards will have one watch holder and a clip for paperwork. The watch holder is reversible for lefthanded technicians. The board is also cut out for the arm and stomach, for comfort's sake.

Three-watch boards are designed to hold the three watches and a common lever for depressing all three crowns at the same time. They also have clips and cut-outs, but lefthanded boards must be special ordered. (See Figure 9-4.)

Digital boards usually have the watches built in. Two watch displays are common. These boards are very expensive. (See Figure 9-7.)

Computer time study does not need a board, just a data collector keyboard.

Videotape Recorder

One of the newest and best tools for studying and recording the method and time standard is the videotape recorder. Operation description is an important part of time study. A time standard is only good for one set of circumstances, and if anything changes, the time standard must change. When a change in time standard is made, it is often challenged by unions because the operation description was not recorded clearly enough. Most unions correctly include a statement like this:

> No time standard (rate) will be changed unless there is a change in the machine, tooling, material, method, or working conditions that creates a change of more than 5%.

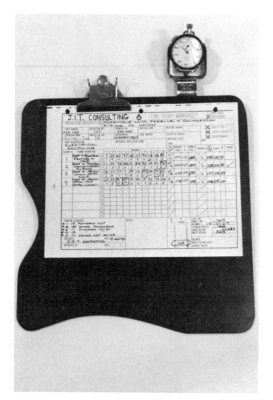

FIGURE 9-11 One-watch time study board. (Courtesy of Meylan Corporation)

Videotape recording a few minutes of operation of a work station only costs a few cents. What better technique is there to record exactly what the time study technician studied than to videotape it?

Another use of videotape is to tape an operation and review the tape for methods analysis and methods improvement. This is called *micromotion study*. The tape can be slowed, speeded, and frozen. It can be replayed to watch one hand at a time. The videotape is a great methods improvement tool.

The third use of videotape recording is to use it as a stopwatch. Many recorders have time recorders built into them. The beginning time can be subtracted from the ending time, and an element or cycle time results. In addition, the time study technician can time-study a tape in the comfort of his or her office. Consider a corporate industrial time study technician going out to a remote plant with a video recorder. One tape can hold fifty operations or more. For a few dollars of tape, the whole plant can be taped and returned to the corporate office for analysis (methods and standards). The corporation will save enough money on travel expenses to pay for the equipment in one trip, plus it will have the best possible record of existing methods.

Every time study department should have a videotape recorder.

FIGURE 9-12 Tachometer: RPM and FPM record. (Courtesy of Meylan Corporation)

Tachometer

A tachometer is used for determining the speeds of machines and conveyers. A center point attachment is placed on the tachometer and then placed against the center of a turning shaft, chuck, or arbor. Revolutions per minute (RPM) is recorded on the time study form as part of the operation description. The center point attachment can be replaced with a 4-inch diameter wheel to convert RPM to feet per minute (FPM) (see Figure 9-12). When the wheel is held against a moving conveyer, FPM is indicated. This information is also an important part of the operation description. The tachometer is an important tool of motion and time study, and without it the technician is left with imperfect information.

Calculator

The importance and use of a calculator does not have to be emphasized to an industrial technologist, but a few comments are needed. Time study involves a lot of math, and the accuracy cannot be overemphasized. The calculator will speed up the process and make the results more accurate.

One special feature is recommended for time study—the 1/x function. 1/x is one divided by another number, and hours per unit and units per hour are the 1/x of each other. The math is much easier and faster, and the time study forms have been set up to be math efficient (i.e., the fewest steps possible). The 1/x button can be one of your favorite functions once you learn its importance and use. Have you ever divided something upside down? (The numerator into the denominator?) If you do it again, just push the 1/x button and it straightens out everything. This is much easier than putting in everything all over again.

Forms

Time study forms are the hardest part of learning how to do a time study. Time study forms are set up to lead the technician into the correct procedure. The next section is a step-by-step procedure for making a stopwatch time study, using a form.

TIME STUDY PROCEDURE AND THE STEP-BY-STEP FORM

The time study procedure has been reduced to ten steps, and the time study form has been designed to help the time study technologist perform the ten steps in the proper sequence. (Figure 9-13 shows a blank time study form with circled numbers.) This section is organized according to the ten sequential steps:

Step 1:   Select the job to study.
Step 2:   Collect information about the job.
Step 3:   Divide the job into elements.
Step 4:   Do the actual time study.
Step 5:   Extend the time study.
Step 6:   Determine the number of cycles to be timed.
Step 7:   Rate, level, and normalize the operator's performance.
Step 8:   Apply allowances.
Step 9:   Check for logic.
Step 10: Publish the time standard.

Within each step, the blocks of the time study form involved are defined. The circled number refers to the block on the time study form. The form is designed for both continuous and snapback time study techniques. Everything except block 16 is exactly the same.

Step 1: Select the job to study. Request for time study can come from every direction:

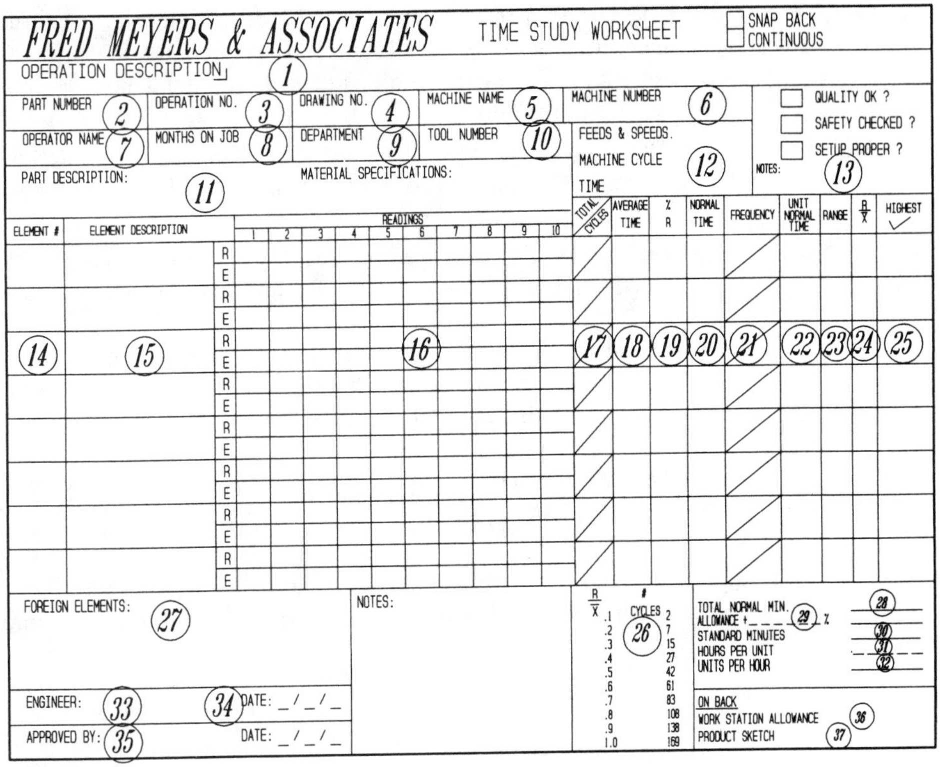

FIGURE 9-13 Time study form: The step-by-step form.

1. Unions can question time standards and request a restudy.

2. Supervisors, who are judged partly on the performance of their people, can request a restudy.

3. The job could change, requiring a new standard.

4. New jobs may have been added to the plant.

5. New products can be added, requiring many new time standards.

6. Industrial technologists can improve methods, requiring a new time standard.

7. Cost reduction programs can require new standards—new machinery, tools, materials, methods, etc.

Once a reason for studying a job has been determined, the time study technician may have several people doing the same job. Which person do you time-study? The best answer is two or three, but those people you do *not* want to time study are

1. The fastest person on the job. The other employees may think you are going to require them to keep up. Even though you can do a good job of setting a time standard on this person, you don't want to create employee relations problems.

2. The slowest person on the job. No matter how you rate the job and no matter how good the time standard is, the employees will wonder how you came up with that standard.

3. Employees with negative attitudes that will affect their performance while being studied. If you can sidestep a potential problem, you should.

     The person or persons to be time-studied should have sufficient time on the job to be a qualified, well-trained operator. For this reason, ⑦ and ⑧ have been included on the time study form:

⑦ Operator's name

⑧ Months on the job.

The employee should have been on the job for at least two weeks.
     Once the job has been selected to study, the following information has been determined:

② Part number

③ Operator number

④ Drawing number

⑤ Machine name: a generic name like press, welder, lathe, drill, etc.

⑥ Machine number: a specific machine with specific speeds and feeds

⑨ Department: where the machine is located (this can be a number or name).

Step 2: Collect information about the job.

     Now that the job has been identified, the technologist must collect information or the purpose of understanding what must be accomplished. The information required is as follows:

① Operation description: a complete description of what needs to be accomplished

④ Drawing number: will lead to a blueprint to show such things as

    a. ⑪ and ㊲ Part description and material specification (a place on the back of the time study form has been set aside for a product sketch, if needed)

   b. ⑩ Tool numbers, and sizes of tools like fixtures, drill sizes, etc.

   c. ⑫ Feeds and speeds of equipment: depend on sizes of parts and material specifications found on the blueprint; they must be recorded.

⑬ When reviewing the work station and before starting the time study, the technologist must check the following:

   √ Is quality OK? Quality control must confirm that the quality of the product being produced is good. Is the operator checking parts on the proper schedule? Time standards from producing scrap are worthless.

   √ Has safety been checked? If all the safety devices are not in place, then the technologist would be wasting time setting a standard for the wrong method.

   √ Is set-up proper? This is the time to see that the proper method, tools, and equipment are in place. Are the materials and tools correctly positioned? Are there unnecessary moves or elements being performed?

If anything is wrong, it must be corrected before a time study can be performed. If the operator must be retrained, the time study should be postponed a couple of weeks.

㊱ A big part of collecting the information is the work station layout. The back of the time study form has been developed for a work station layout, but this may not be needed if done on another one of the previous forms (multi activity form). The work station layout is one of the best ways to describe the operation. Review chapter 7, on work station design, for what must be included on a work station layout.

Step 3:   Divide the job into elements.

   Elements were discussed in chapter 8 as being a unit of work that is indivisible. This is not a useful definition in time study, because the operation could be too fast to time-study. Time study elements should be as small as possible, but not less than .030 minutes.
   The element should be as descriptive as possible. The elements must be in the sequence that the methods call for and should be broken down as small as is practical.

Principles of Elemental Breakdown

1. It is better to have too many elements than too few.
2. Elements should be as short as possible, but not less than .030 minutes. Elements over .200 minutes should be examined for further subdivision.

3. Elements that end in sound are easier to time because the eyes can be looking at the watch while the ears are anticipating the sound.

4. Constant elements should be segregated from variable elements to show a truer time.

5. Separate the machine-controlled elements from the operator-controlled elements so work pace can be differentiated.

6. Natural breaking points are best. The beginning and ending points must be recognizable and easily described. If the element description isn't clear, the description or breakdown must be rethought.

7. The element description describes the complete job, and the ending points are clearly marked.

8. Foreign elements should be listed in the order of occurrence. Foreign elements are not listed until they occur during the study.

The reasons for breaking down a job into elements are as follows:

1. To describe the job.

2. Different parts of the job have different tempos. The time study technician will be able to rate the operator better. Machine-controlled elements will be constant and normally 100%, whereas the operator may be more or less proficient at different parts of the job.

3. Breaking down the job into elements allows for moving a part of the job from operator to operator. This is called *line balancing*.

4. Standard data can be more accurate and more universally applied with smaller elements. All work is made up of common elements. After a number of time studies, the technologist can develop formulas or graphs to eliminate the need for time study. Standard data is the goal of all time study departments.

On our time study form, two columns have been assigned to elements:

⑭ Element #: The element number is just a sequential number and is useful when more than ten cycles are timed. Instead of describing each element over and over again, we just reference the element number.

⑮ Element description: Be as complete as possible. The ending points should be clear.

㉗ Foreign elements: These foreign elements will be eliminated from the study, but we don't want to hide anything. Therefore, a reason for throwing out the time is required. Foreign elements marked with an asterisk (*) in the body of the study are referred to this box.

Step 4:　　The actual time study: ⑯

This is the guts of the stopwatch time study. Block 16 on the step-by-step form is for recording the time for each element. The form has room for eight elements (eight lines) and ten cycles (columns) for eighty readings. Most studies will have only three or four elements, so there is room on one sheet for twenty cycles. This form can be used for either snapback or continuous time study.

Continuous time study is the most desirable time study technique. The stopwatch remains running through the duration of the study, and element ending times are recorded behind the "R" for reading.

CONTINUOUS EXAMPLE

		1	*2*	*3*	*4*	*5*
	R	16	.83	1.50	2.17	2.83
Load and clamp	E					
	R	56	1.23	1.90	2.57	3.23
Run machine	E					
	R	66	1.33	2.01	2.67	3.32
Unload & aside	E					

Note that each time is getting larger and that five parts were run in a total time of 3.32 minutes. In step 5, we calculate the elemental times, but at this time we are still out in the plant collecting data.

Snapback studies allow the technician to read the watch and reset it immediately to time the next element. The exact same study is shown next using the snapback technique.

SNAPBACK EXAMPLE

		1	*2*	*3*	*4*	*5*
	R					
Load & clamp	E	.16	.17	.17	.16	.16
	R					
Run machine	E	.40	.40	.40	.40	.40
	R					
Unload & aside	E	.10	.10	.11	.10	.09

Note that the elemental time (E) is already calculated. Look at the load and clamp time; the times look consistent—16, 17, 17, 16, and 16. The time for loading and clamping is immediately obvious. This same information will be available in a continuous time study, but a lot of arithmetic is required first. In the snapback time study technique, the "R" row can be used for rating the operator on each element of work. (We discuss this in more detail later when we discuss the rating, leveling, and normalizing sections.)

Step 5:   Extend the time study.

Now that the time study has been taken, the bigger job comes. The continuous method has one more step than the snapback method, so we will concentrate on the continuous method. After a brief description of the steps, an example problem is given for you to extend. (See Figure 9-18.)

⑯ Subtract the previous reading from each reading. The previous element reading was its ending time and the beginning of this element. Subtracting the beginning time from the ending time gives elemental time.

⑰ Total/cycles: The total refers to the total time of the good cycles timed. Some cycles may be eliminated because they include something that doesn't reflect the elemental time.

   Foreign elements are eliminated from further consideration. Cycles are the number of good elemental times included in the total time.

⑱ Average time: Average time is the result of dividing total time by the number of cycles. On the average, it took .40 minutes machine time on our last example.

⑲ % R: Percent rating refers to our opinion of how fast the operator was performing. The rating divided by 100, multiplied by the average time, equals normal time.

$$\text{Average time} \times \frac{\text{rating }\%}{100} = \text{normal time.}$$

Later in this chapter we discuss rating in detail.

⑳ Normal time: Normal time is defined as the amount of time a normal operator working at a comfortable pace would take to produce a part. Normal time is calculated above and is explained further in step ㉒.

㉑ Frequency: Frequency indicates how often a task is performed. For example, moving 1,000 parts out of the work station, moving the empty tub to the other side of the work station, and bringing in a full tub of 1,000 new parts to the work station would occur only once in 1,000 cycles (1/1,000). If Quality Control asked the operator to inspect one part out of every 10, $\frac{1}{10}$ would be placed in this column. The biggest use of this column is when the operator is doing two parts at a time; then $\frac{1}{2}$ is placed in this column. If $\frac{1}{1}$ goes in the column, it can be left blank.

㉒ Unit normal time: Unit normal time is calculated by multiplying the frequency by the normal time.

Examples:

Normal Time		Frequency		Unit Normal Time
1.160	×	1/1,000	=	.001 minutes
.400	×	1/10	=	.040 minutes
.100	×	1/2	=	.050 minutes
.050	×	1.1	=	.050 minutes

Every element must reflect the time to produce one unit of production. No one wants a standard for pairs, and mixing frequency of units leads to bad time standards. Be very careful here.

Step 6: Determine the number of cycles to be timed.

The accuracy of time study is dependent on the number of cycles timed. The more cycles studied, the more accurate the study. Almost all time study work is aimed at an accuracy of ±5% with a 95% confidence level, so the question turns to how many cycles to study to achieve this accuracy.

Graphs and tables are easier and more cost effective than formulas. The graphs and tables (Figures 9-14 & 9-15) used in this book are based on the formula

$$N = \sqrt{N} = \frac{2R}{Ad_2\bar{x}} \quad \text{or} \quad \frac{4R^2}{(A)^2(d_2)^2(\bar{x})^2}$$

R = Range of the sample of observations
(highest value of elemental data minus lowest value)

A = Required precision (±5% or ±10%)
(must be extended as decimal ±.05 or ±.10 etc.)

d_2 = A constant which is used to estimate the standard deviation of a sample and is a function of the sample size. Must be obtained for a statistical table.

$\bar{x}$ = Arithmetical average; sum of the observations divided by the number of observations.

Example: Confidence 95% precision ±5%
Readings .08, .07, .09, .10, .07, .11, .08, .07
Range = .11 − .07 = .04
Required precision = .05
d_2 = 3.078
$\bar{x}$ = .82/10 = .082

$$N = \frac{4(.04)^2}{(.05)^2(3.078)^2(.082)^2} = 40.18 \text{ or } 40$$

Second Example: Everything the same except the precision (A) = ± 10% (.1).

$$N = \frac{4(.04)^2}{(.1)^2(3.078)^2(.082)^2} = 10.01 \text{ or } 10$$

If ±10% accuracy is an acceptable quality standard for our time study data, one fourth the number of cycles are required (40 vs. 10).

Figure 9-14 shows two graphs as an example of how many cycles to time. These two graphs are based on an initial ten or twenty timed cycles. Calculate the range *(R)* and the arithemetic average $\bar{x}$. Divide *R* by $\bar{x}$ and *A* factor results. Look up the factor on either chart. The bottom chart is used when the factor is under 1.0, and the top chart is used when the factor is over 1.0. The charts tell you how many cycles to time to achieve ±5% accuracy. Divide this number by 4 if ±10% accuracy is adequate.

The number of cycles to time-study is a measure of how confident we are in the time standard we are setting. The time standard is much like a batting average or a bowler's average. In the beginning of a season, a good or bad game can change the average considerably. At the end of a season, however, a good or bad game will have very little effect on the average. We want to take enough cycle readings to ensure that a good or bad cycle will not affect the average. We can then have the confidence that our time standard is achievable and a fair reflection of what the operators can do.

The following blocks on time study form help calculate the number of cycles needed:

㉓ The range

㉔ $R/\bar{x}$ or the factor

㉕ Highest

㉖ Factor table.

The procedure for determining the number of cycles to be timed is as follows:

1. Time-study ten cycles for jobs less than 2 minutes long and five cycles if longer than 2 minutes.
2. Determine the range *(R)* ㉓ of the elemental times for each element of the job. The range is the highest elemental time less the lowest elemental time. The smaller this range, the fewer cycles needed for the accuracy level.
3. The average time has already been determined in column ⑱ of the time study. $\bar{x}$ is the mathematical symbol for arithmetic average.

An example of one element of a job is: .08, .09, .08, .08, .07, .08, .08, .10, .08, .09. The total of these elemental times is .83; the number of elemen-

FIGURE 9-14

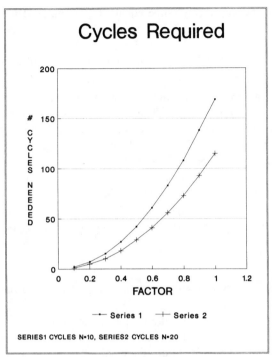

Cycles Required

SERIES1 CYCLES N=10, SERIES2 CYCLES N=20

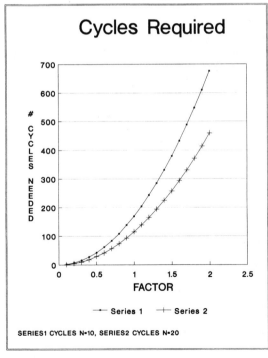

Cycles Required

SERIES1 CYCLES N=10, SERIES2 CYCLES N=20

tal times is 10; therefore, average time equals .083 minutes. The range *(R)* is
.10 − .07 = .03.

4. Determine the factor $R/\bar{x}$ 24; $R/\bar{x}$ is the range divided by the average time.
   $R = .03$, $\bar{x} = .083$, and $R/\bar{x} = .36$.

5. Determine the number of cycles needed. ㉖ on the time study form is a list
   of factors most commonly used. The number .36 on the table is 60% of
   the way between .3 and .4. Sixty percent of the difference between 27
   cycles and 15 cycles is 7.2 cycles; therefore, 23 cycles must be timed
   (7.2 + 15 = 22.2). The time study technologist must go back to the job and
   collect 13 more cycles to have a 95 ± 5% accurate time study (see Figure
   9-15). Factors over 1.0 are possible, as the graphs in Figure 9-14 show,
   but the cost would be very high. Something is wrong if you have a factor
   over 1.0—probably an inexperienced operator or time study person.

Normal Distribution

The number of cycles to time study is based on the laws of probability as best explained
by the normal distribution curve.

   If we were to collect data on any specific characteristic of a large group of individuals, or if we were to measure a particular dimension on a large number of parts,
and then plot the frequency of each observation, the resulting distribution would resem-

FIGURE 9-15 The number of cycles to
time based on 95% ± 5% accuracy

$\dfrac{Range = R}{Average\ Time = \bar{x}}$	No. Cycles to Time
.1	2
.2	7
.3	15
.4	27
.5	42
.6	61
.7	83
.8	108
.9	138
1.0	169

Notes:
1. If ± 10% accuracy is OK, divide the number of cycles by 4.
2. A result over 169 means that you have timed an
   unskilled operator, or your readings are off, or your
   elements are too small.
3. This chart is printed at the bottom of the time study
   forms in this book.

ble a symmetrical bell-shaped curve. Such distributions are called *normal* or *Gaussian distributions* and are of significant importance in any statistical study. Although there are many other forms of distributions, most phenomena in nature and industry have a normal distribution; or most often these distributions approach normality to such an extent that the normal laws of probability are applicable.

Understanding normal distributions is simple. In such cases, most values are bunched or clustered around the central value or the average. As the values differ from the average value, and as the size of this difference increases, the frequency or occurrence of these values deceases. Consider an employee who, on the average, takes 15 minutes to perform a task. We know from experience that sometimes he or she will take a little more, and sometimes a little less time to do the same task. However, seldom does the employee take too much time under or over the average time. Over a long period of time, the employee will have an approximately equal number of extremely low and high readings, although not very many of each.

Standard Deviation

Given any normal distribution, whereas the average indicates around what value most of the data are clustered, it does not indicate how variable or spread out the data may be. Imagine two groups of employees performing the same task (Figure 9-16). One

FIGURE 9-16 Normal distribution curve.

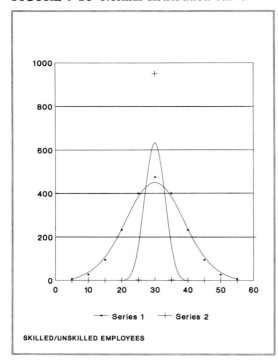

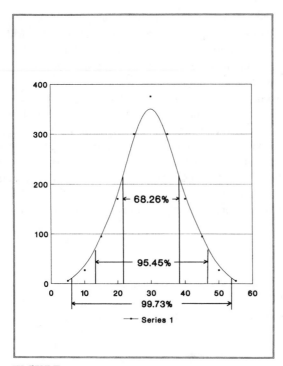

FIGURE 9-17 Standard deviation curve.

group consists of individuals with an equal amount of training and experience. The employees in the second group, however, are varied in training and experience. While the *average* time per employee for both groups may be the same (say 30 minutes), the individual times in the first group may range from 25 to 35 minutes; and in the second group from 10 to 50 minutes. Would you say that both distributions are the same? Of course not. Even though both may be normal and have the same average, they do not have the same spread or variability. The quantitative value which states the degree of variability or spread of a population is called *standard deviation* and is denoted by *s*. The larger the variability or the spread of the data, the larger the standard deviation.

There is a definite and interesting relationship between the variability of the data as measured by the standard deviation and the frequency or occurrence of the values for any distribution. Given the average and the standard deviation, *s*, approximately 68.26% of all the measurements for any normal distribution will be located between the average ±1*s*. (See Figure 9-17.) If the average is extended to ±2*s* on both sides (i.e., average ±2*s*), then 95.45% of all the data are included. Finally, the area which is enclosed by average ±3*s* will include approximately 99.73% of the data.

Step 7:   Rate, level, and normalize the operator's performance.

⑲ % rating is the technologist's opinion of the operator's performance. Rating, leveling, and normalizing all mean the same thing, and the term *rating* is used from this point on. Rating is the most challenged aspect of motion and time study, and for that reason the most important subject of this chapter (it is discussed in detail later in this chapter).

$$\text{(Average time)} \times \frac{\text{(rating)}}{100} = \text{normal time}$$
$$\qquad\quad ⑱ \qquad\qquad\quad ⑲ \qquad\qquad\quad ⑳$$

Step 8:   Apply allowances ㉙

Allowances are added to a time study to make the time standard practical.

$$\text{(Total normal time)} + \text{(allowances)} = \text{standard time}$$
$$\qquad\quad ㉘ \qquad\qquad\qquad ㉙ \qquad\qquad\qquad ㉚$$

There are several methods of applying allowances, and there are several types of allowances. We discuss allowances in detail later in this chapter.

Step 9:   Check for logic.

Once the time study has been extended, the test for logic should be applied in two ways:

1. The average time ⑱ should look like the elemental times. If an error in adding was made, a test for logic will save a mistake. The easiest mistake to make is decimal error. Be careful not to make decimal errors, because they look really bad—1,000% errors result from misplacing a decimal only one place. That is why being consistent with decimal placement is so important:

   a. Read stopwatches in two places: .01

   b. From average time on, use three places: .001

   c. Hours per unit are five places: .00001.

2. The second test for logic is the total normal time for one unit. During your study, you timed a specific number of parts in a certain amount of time. For example, ten cycles were timed in 7.5 minutes (7.5 was the last reading in the tenth column). The average time should be somewhere around .75 minutes each. Are you close with the total normal time? If not, there is a major error. Warning: Don't forget that if the operator is producing two at a time, twice as many parts are being produced.

Step 10: Publish the time standard.

Three numbers are required to communicate a time standard:

1. Standard minutes: ㉚
2. Hours per unit: ㉛
3. Pieces per hour: ㉜.

Starting with standard minutes, dividing ㉚ by 60 minutes per hour equals hours per unit, ㉛ and pieces per hour ㉜ is $1/x$ of ㉛ (or divide hours per unit into 1 hour).

Every company has a method of recording time standard information. In chapter 3, an operation sheet for a water valve factory was shown. The time standards could be placed on that operations sheet (please review Figure 3-1). The production route sheet is another common tool for communicating the time standard. The computer is the most common method of storing and communicating to everyone what the time standard is for each job.

A few more pieces of information remain to be discussed on the step-by-step time study form:

㉝ Engineer: The time study technologist puts his or her name here.

㉞ Date: A time study with an incomplete date is worthless.

㉟ Approved by: This is where the chief engineer or manager signs, approving your work. You never fill this in.

An example problem has been included in Figure 9-18. The data has been collected, the job broken down into elements, and the time study has been made. You need to extend the study and develop a time standard. This was a continuous time study, and that should be obvious because the times are always getting larger. The extension will start with the subtraction of element readings to find elemental time.

RATING, LEVELING, AND NORMALIZING

Rating is the process of adjusting the time taken by an individual operator to what could be expected from a normal operator. The industrial technologist must understand the industry standards of *normal*.

Rating an operator includes four factors:

1. Skill
2. Consistency

FRED MEYERS & ASSOCIATES TIME STUDY WORKSHEET

☐ SNAP BACK
☐ CONTINUOUS

OPERATION DESCRIPTION: ASSEMBLE PARTS 2 & 4, MACHINE SCREW & STAKE. INSPECT

PART NUMBER 4650-0950	OPERATION NO. 1515		DRAWING NO. 4650-0950	MACHINE NAME	MACHINE NUMBER 21			☐ QUALITY OK ?
OPERATOR NAME MEYERS	MONTHS ON JOB 5		DEPARTMENT ASSEMBLY	TOOL NUMBER M61	FEEDS & SPEEDS.			☐ SAFETY CHECKED ?
PART DESCRIPTION:			MATERIAL SPECIFICATIONS:		MACHINE CYCLE TIME		NOTES:	☐ SETUP PROPER ?

ELEMENT #	ELEMENT DESCRIPTION		1	2	3	4	5	6	7	8	9	10	TOTAL CYCLES	AVERAGE TIME	% R	NORMAL TIME	FREQUENCY	UNIT NORMAL TIME	RANGE	R/X	HIGHEST ✓
	ASSEMBLY	R	9	41	71	1.07	38	77	2.08	48	77	3.07		90			1	1			
		E																			
	DRIVE SCREW	R	15	46	79	13	43	82	14	53	82	93		100			1	1			
		E																			
	PRESS	R	28	59	94	27	66	95	28	66	96	4.06		110			1	1			
		E																			
	INSPECT	R	32	62	92	30	69	98	41	69	99	4.09		100			1	1			
		E																			
	LOAD SCREWS	R										3.83		125			1	10			
		E																			

FOREIGN ELEMENTS:		NOTES:	R/X .1 .2 .3 .4 .5 .6 .7 .8 .9 1.0	# CYCLES 2 7 15 27 42 61 83 108 138 169	TOTAL NORMAL MIN. ALLOWANCE +____ 10 % _____ STANDARD MINUTES HOURS PER UNIT - - - - - UNITS PER HOUR _____
ENGINEER:	DATE: _ / _ / _				ON BACK WORK STATION LAYOUT
APPROVED BY:	DATE: _ / _ / _				PRODUCT SKETCH

FIGURE 9-18 Time study problem: Continuous technique

3. Working conditions
4. Effort (most important).

Three of these four factors are accounted for in other ways and have little effect on rating. Effort will be our primary concern.

1. *Skill:* The effect of skill is minimized by timing only people who are skilled. Operators must be fully trained in their work classification before being time-studied. A welder must be a qualified welder before being considered a subject for time study. Two years of training may be required to become a welder, and in addition, this welder must be on this job for at least two weeks before performing the job sufficiently. Habits of motion patterns must be routine enough that the operator doesn't have to think about what comes next and where everything is located. Very skilled operators make a job look easy, and the industrial technologist must let this skill affect the rating. On the other hand, if an operator shows lack of skill, such as dropping, fumbling, inconsistent times, stopping/starting, etc., the technologist should postpone the study or find someone else to time-study.

2. *Consistency:* Consistency is the greatest indication of skill. The operator is consistent when he or she runs the elements of the job in the same time, cycle after cycle. The time study technician begins to anticipate the ending point while looking at the watch and listening for the ending point. The operator is said to be like a machine. Consistency is used to determine the number of cycles. A consistent operator needs only to run a few parts before the cycle time is known with accuracy. The skill of the operator should be evident to the time study technician, and the technician's rating of the operator should be high. When inconsistency is present, the technologist must take many more cycles to be acceptably accurate in the time study. This inconsistency tends to affect the technologist's attitude and rating of the operator in a negative way, and the best thing to do is find someone else to study. It is more fun rating and working with operators with great skill.

3. *Working conditions:* Working conditions can affect the performance of an operator. In the early twentieth century, this was much more of a problem than today, but if employees are asked to work in hot, cold, dusty, dirty, noisy environments their performance is going to suffer. These poor working conditions can be eliminated if the true cost is shown. The way we account for poor working conditions today is to increase the allowances (discussed later in this chapter). If operators are required to lift heavy materials in the performance of their duties, 25% more time can be added to the time standard as an allowance. Working conditions are not a part of modern rating.

4. *Effort:* Effort is the most important factor in rating. Effort is the operator's speed and/or tempo and is measured based on the normal operator working at 100%. A 100% performance rating is defined as

a. Walking 264 feet in 1.000 minutes or 3 miles per hour

b. Dealing fifty-two cards into four hands around a $30'' \times 30''$ card table in .500 minutes.

c. Assembling thirty $\frac{3}{8}'' \times 2''$ pins into a pinboard in .435 minutes.

Effort can be easily seen in walking. Walking at speeds less than 100% is uncomfortable for most people, and walking at 120% requires a sense of urgency that indicates increased effort.

Psychology has been good to the time study technician. The normal tendency of people being watched is to speed up. Being watched makes people nervous, and nervous energy is converted by the body into a faster tempo. The time study technician then gets a frequent chance to rate over 100%. When an operator works at 120%, the technologist has the pleasant experience of telling the operator, "You are fast. I'm going to have to give you 20% more time so that an average person can do the job." That is fun to say, and it happens often.

When rating, you must keep tuned into normal pace. This requires continued practice on your part, forever. The experiments in the next section will help keep your rating accurate.

PTSS has been developed based on the concept of normality according to industry standards, and a synthetic rating is developed by time-studying a job that has been proven by PTSS. A good learning technique used at many companies is to have new technologists time-study known jobs and compare their time standards to the known time standards. Another good learning experience is to time several people on the same job. Effort and skill are the only differences in time, so proper rating should make all the normal times the same.

Many companies use time study rating films developed by industrial associations and professional organizations:

1. Society for the Advancement of Management (SAM)
2. Tampa Manufacturing Institute
3. Ralph Barnes and Associates.

All of these groups produce time study rating films. *Industrial Engineering* magazine would be a good source as well.

100% STANDARDS AND EXPERIMENTS

Industrial technologists can teach themselves to be good raters by setting up some simple experiments.

Walk a 50-foot Course

Set up a 50-foot course with a starting line and an ending line. With a stopwatch in hand, start about 10 feet in front of the starting line and start the walk. When crossing the starting line (already up to speed), start the stopwatch. Maintain the same pace until the finished line is crossed; then stop the watch. Slow down and stop only after the finish line is crossed. Now read the watch. Let's use an example of .18 minutes.

How fast was the walk? The time should have been .19 minutes (the standard). The time standard for walking has been universally accepted as 3 miles per hour.

$$\frac{3 \text{ miles per hour} \times 5280 \text{ ft./mile}}{60 \text{ minutes/hour}} = 264 \text{ ft./minutes}$$

$$\frac{50 \text{ ft.}}{264 \text{ ft./min.}} = .19 \text{ minutes [time standard for 50 feet].}$$

Notice that only two decimal places are used when reading a stopwatch.

$$\% \text{ Performance} = \frac{\text{time standard}}{\text{actual time}} = \frac{.19}{.18} = 105\%$$

What is the percent performance for the following?

Stopwatch Time (Actual)	Standard Time	% Performance
.24	.19	
.16	.19	
.28	.19	
.11	.19	
.20	.19	

FIGURE 9-19 RATING: Percentages and time for walking a 50-foot course

Time	%	Time	%	Time	%
.09	211	.18	106	.27	70
.10	190	.19	100	.28	68
.11	173	.20	95	.29	66
.12	158	.21	90	.30	63
.13	146	.22	86	.31	61
.14	136	.23	83	.32	59
.15	127	.24	79	.33	58
.16	119	.25	76	.34	56
.17	112	.26	73	.35	54

A test of logic is, did you walk faster (over 100%) or slower (under 100%) than the standard?

Practice walking the 50-foot course at 100% first. Then try 120%, 80%, and 60%. Notice how difficult walking below 100% is. Notice the sense of urgency at 120%.

Walking is a good first study of rating. Most everyone is a skilled walker, so effort is all that is being rated. Some people are tall and have long legs, while other people are short and have short legs. Is it fair to have one standard for all? The answer doesn't speak to the fairness, but only to the fact that one standard is all that can be used. Would a customer be willing to pay more because a short person worked on this job? No. Only one cost can be charged, and only one time standard can be used. The "average" person must be used in all time study. Figure 9-19 shows time and percentages for walking a 50-foot course.

Dealing Fifty-two Cards Into Four Equal Stacks On A Card Table In .500 Minutes

Set up an area to deal cards. A 30-inch square card table was used, but any area of at least that size can be used. A good start would be to start a stopwatch and place it in front of the dealer, who then deals a card every .01 minutes. When fifty-two cards are dealt, .52 minutes have been used. What is that performance?

$$\text{Performance} = \frac{\text{time standard}}{\text{actual time}} = \frac{.50}{.52} = 96\%$$

Ninety-six percent is close to 100%, and if a small speed-up were to be made, 100% would result. Dealing cards is something that we all do very well. Again, give it a try. Deal 100% a few times, then 120%, 80%, etc. Watch other people and estimate their rating without the use of the stopwatch. Check yourself with the watch, but don't let it affect your rating estimate.

What are the percent performances for the following?

Dealing Time	Standard	% Performance
.40	.50	
.45	.50	
.55	.50	
.60	.50	

Try dealing the cards as fast as you can without losing pile quality. Try dealing the cards as slowly as you can without stopping between cards. The range will probably be a ratio of 1:2, the slowest time takes twice as long as the fastest.

Dealing cards is a good rater training experiment because it is much like simple assembly work. Reaches and moves account for half of all work and are the only thing a technologist can rate. Dealing cards consists of reaches and moves. Figure 9-20 is times and percentages for dealing 52 cards into 4 equal stacks.

FIGURE 9-20 Rating: Percentages and times for fifty-two dealing cards (four equal stacks).

Time	%	Time	%	Time	%
.25	200	.46	109	.67	75
.26	192	.47	106	.68	74
.27	185	.48	104	.69	72
.28	179	.49	102	.70	71
.29	172	.50	100	.71	70
.30	167	.51	98	.72	69
.31	161	.52	96	.73	68
.32	156	.53	94	.74	68
.33	152	.54	93	.75	67
.34	147	.55	91	.76	66
.35	143	.56	89	.77	65
.36	139	.57	88	.78	64
.37	135	.58	86	.79	63
.38	132	.59	85	.80	63
.39	128	.60	83	.81	62
.40	125	.61	82	.82	61
.41	122	.62	81	.83	60
.42	119	.63	79	.84	60
.43	116	.64	78	.85	59
.44	114	.65	77	.86	58
.45	111	.66	76	.87	57

Assemble Thirty Pins into a Pinboard in .435 Minutes Using the Two-hand Method

The assembly of the pinboard experiment has been around for almost 100 years. The PTSS time is .435 minutes. The two hands are working at the same time; therefore fifteen sets of motions are required. The pinboard experiment is realistic and similar to production work. It has been used in the past to screen new employees. Figure 9-21 is a picture of the 30 pin pinboard assembly job, and Figure 9-22 is the times and percentages for assemblying the pinboard.

What would be the percent performance of the following pinboard times?

	Stopwatch Time in Minutes	Standard	% Performance
1.	.32		
2.	.40		
3.	.50		
4.	.45		

Rating is not an exact science. The industrial technologist should aim for ±5% accuracy, and ±10% accuracy is of value. When rating, always round off to the nearest 5% (80, 85, 90, 95, etc.). Anything else is a false claim to more ability than we possess.

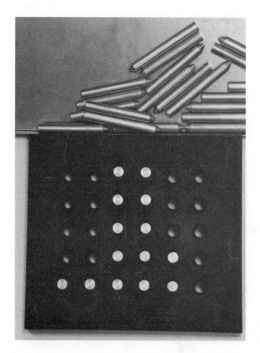

FIGURE 9-21 1. The pinboard is 7″ × 8″ with five rows of six pins, $\frac{3}{8}$″ diameter. The holes are countersunk on one side. 2. The thirty pins are $\frac{3}{8}$″ diameter, $2\frac{3}{4}$ inches long, beveled on one end. 3. Assemble thirty pins using two hands. Countersink up in .435 minute.

FIGURE 9-22 Ratings: Percentages and times for assembling pinboard (thirty pins, two hands).

Time	%	Time	%	Time	%
.27	161	.43	101	.59	74
.28	155	.44	99	.60	73
.29	150	.45	97	.61	71
.30	145	.46	95	.62	70
.31	140	.47	93	.63	69
.32	136	.48	91	.64	68
.33	132	.49	89	.65	67
.34	128	.50	87	.66	66
.35	124	.51	85	.67	65
.36	121	.52	84	.68	64
.37	118	.53	82	.69	63
.38	114	.54	81	.70	62
.39	112	.55	79	.71	61
.40	109	.56	78	.72	60
.41	106	.57	76	.73	60
.42	104	.58	75	.74	59

TIME STUDY RATER TRAINER FORM

The time study rater trainer form is an aid to help the time study technician improve his or her rating ability. The form has been developed to point out the technician's problems and to give direction to improvement. The step-by-step procedure that follows is in conjunction with the form, which has circled numbers on it (see Figure 9-23). The step-by-step procedure will tell the technician exactly how to use the form. The experiments just discussed (walking, dealing cards, or pinboard) or rating films can be used to test the technologist's rating ability, but this form is used to test the technologist's ability to rate the performance of any operation. Ten rating samples are needed and are compared to the actual time.

STEP-BY-STEP PROCEDURE FOR COMPLETING THE TIME STUDY RATER TRAINER FORM

①. Name: Your name goes here—the person doing the rating. This form is a measure of rating ability, so the person doing the rating is the focus of this form.

②. Date: The full date of the test.

③. Task: Walking, dealing cards, pinboard, film name or number, or job title. What job did you watch?

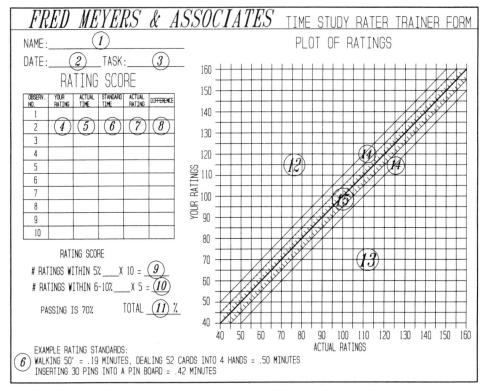

FIGURE 9-23 Time study rater trainer form: The step-by-step form

④. Your rating: There are ten rows in this column, and one rating goes in each column. Look at the completed example for walking (Figure 9-24). When ten people were observed (one at a time) walking through a 50-foot course, the rating for each of these people was placed in this column. This column is completed before any further work is done.

⑤. Actual time: While observing each person walking through the 50-foot course, the technologist should time them with a stopwatch. This time is recorded in this column. This is the actual time it took the person to walk through the 50-foot course. Don't let the watch reading affect your rating.

⑥. Standard time: The standard time is

a. Walking = .19 minutes

b. Dealing cards = .500 minutes

c. Pinboards = .435 minutes

d. Films will give you actual ratings ⑦.

Standard time is the time a skilled person would take if working at a normal pace. The standard time will be the same for each observation, as long as the task is the same.

⑦. Actual rating: The actual rating is

$$\frac{⑥ \text{ Standard time}}{⑤ \text{ Actual time}} = \text{actual rating ⑦.}$$

The actual rating is the correct answer that the technologist was shooting for.

⑧. Difference:

Actual rating ⑦ − your rating ④ = difference ⑧

or

Your rating ④ − actual rating ⑦ = difference ⑧

(being over or under makes no difference here).

⑨. Ratings within 5% _____ × 10 = _____.

Ratings with a difference of 5% or less are perfect ratings and for grading earn ten points each. If a technologist rated all ten observations within ±5%, a 100% grade would result $(10 \times 10 = 100)$.

⑩. Ratings within 6–10% _____ × 5 = _____.

Ratings with a difference ⑧ of less than 10%, but more than 5%, are good ratings and are worth five points each on a grading scale. If a technologist rated all ten observations with 6–10%, a 50% grade would result $(10 \times 5 = 50)$. Fifty percent isn't great, but zero percent is worse. As can be seen, ratings of over 10% difference ⑧ earn no grade points.

⑪. Total _____%

The total percent is calculated by adding the results of ⑨ + ⑩ = ⑪. Just like class grades, 60% is passing, 70% = C, 80% = B, and 90% = A. A technologist working with motion and time study must maintain an 80% plus rating skill grade.

A comment must be made for the ± aspect of rating. The ± gives the technologist an edge on rating; a ±10% error will be offset by a −10% error, and the total will be correct. Errors will have a chance of washing out if the technologist is zeroed in at the beginning.

⑫. Plot of rating (loose) above the line

The plot of rating allows the technologist a chance to see his or her rating ability. The technologist takes each pair of ratings (his or her rating ④ and the actual rating ⑦) and plots them on a graph. His or her rating ④ goes on the vertical axis, and the actual rating ⑦ on the horizontal axis. Where the two points intercept on the graph, place an X. The ⑫ area on the graph indicates that the technologist is too loose or that the technologist rates too high. If many of the ratings are too high, the technologist should consciously lower the rating.

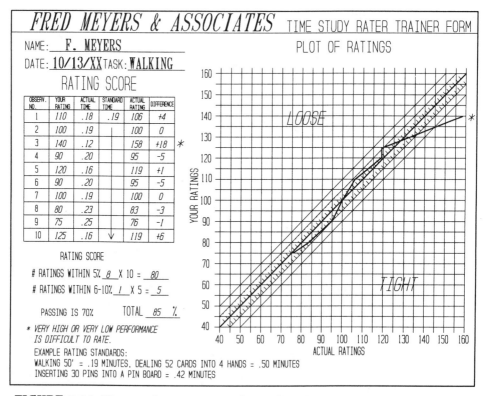

FIGURE 9-24 Time study rater trainer form: Completed example

⑬. Plot of rating (tight) under the line

Plotted points under the line indicate that the technologist is rating too low or is rating tight. Many ratings in this area indicate that the ratings should be increased.

⑭. Between the ±10% line

This is good rating—not great, but good.

⑮. Between the ±5% line

This is the goal line. In rating, the line is ±5% and a rating within ±5% is great rating. This is the goal of all our ratings.

Once all the points are plotted, connect the X's by drawing a straight line between them. (See the example in Figure 9-24).

Adjust your thinking, if needed, and try again. Your rating score needs to be 80% plus.

Figure 9-25 is a list of ten fundamentals of pace rating provided by the Tampa Manufacturing Institute. Figure 9-25 is the back side of the time study rater trainer form 9-24.

1. Healthy people in the right frame of mind easily turn in 100% performance on correctly standardized jobs. For incentive pay, good performers usually work at paces from 115% to 135%, depending on jobs and individuals.
2. For most individuals, *it is uncomfortable to work at a tempo much below 100% and extremely tiring to operate for sustained periods at paces lower than 75%;* our reflexes are naturally geared to move faster.
3. Poor efficiency on a correctly standardized job usually results from stopping work frequently—"goofing off" for a variety of reasons. Specifically, *substandard production seldom results from inability to work at a normal pace.*
4. Some standards of 100%   1. Walking 3 m.p.h. or 264 ft./min.
                            2. Dealing cards into 4 stacks in .5 min.
                            3. Filling the pinboard in .435 min.
5. Very seldom can *true* performance of over 140% be found in industry.
6. Where an operator consistently comes up with extremely high efficiencies, it is usually a sign that the method has been changed or the standard was wrong in the first place.
7. An operator's work pace during a time study does not affect the final standard. His or her *actual* time is multiplied by the performance rule to give a job standard which is fair for all employees.
8. Inasmuch as healthy employees can *easily* vary work pace from approximately 80% to around 130%—through a range of 50%—reasonable inaccuracies in the setting of standards should be sensibly accepted.
9. Ineffective foremen usually fight job standards. *Good supervisors, however, sincerely help in the standard-setting effort, clearly realizing that such information is their best planning and control tool.*
10. *Methods usually influence production more than work pace.* Don't ever get so absorbed in how quickly or how slowly an operator "seems" to be moving that you fail to consider whether or not he or she is using the right method.

FIGURE 9-25 Ten fundamentals of pace rating. (Courtesy of Tampa Manufacturing Institute)

ALLOWANCES AND FOREIGN ELEMENTS

Allowances are extra time added to the normal time to make the time standard practical and attainable. No manager or supervisor expects employees to work every minute of the hour. What should be expected of the employee? This was the question asked by Frederick W. Taylor over 100 years ago. Would you expect the employee to work 30 minutes per hour? How about 40 minutes? 50 minutes? This section will assist the technologist in answering that question.

Types of Allowances

Allowances fall into three categories:

1. Personal
2. Fatigue
3. Delays.

Personal Allowance. Is that time an employee is allowed for personal things such as

a. Talking to friends about nonwork subjects;
b. Going to the bathroom;
c. Getting a drink;
d. Or any other operator-controlled reason for not working.

People need personal time, and no manager would begrudge an appropriate amount of time spent on these activities. An appropriate amount of time has been defined as about 5% of the work day, or 24 minutes per day.

Fatigue Allowance. Is the time an employee is allowed for recuperation from fatigue. Fatigue allowance time is given to employees in the form of work breaks, more commonly known as coffee breaks. Breaks are of varying intervals and varying duration, but all breaks are designed to allow employees to recuperate from the fatigue their jobs produce. Most employees today have very little physical drudgery involved with their jobs, but mental fatigue is just as tiring. If an employee uses less than 10 pounds of effort during the operation of his or her job, then 5% fatigue allowance is normal. A 5% increase in fatigue allowance is given for every 10-pound increase in exertion required of the employee (see Figure 9-26).

Example:

An employee must pick up a 50-pound part.
Fatigue allowance is $50 - 10 \div 10 = 4.0$ units of ten pounds.

$$5\% + (4 \times 5) = 25\% \text{ allowance}$$

Explanation of example

The basic fatigue allowance is 5% and an additional 5% fatigue allowance is added for each 10 pounds of force required over 10 pounds. 50 pounds is 40 pounds more than basic. 40 pounds is 4 units of excess weight (10 pounds is one unit). 4 units × 5% = 20% of excess fatigue. 20% + 5% basic equals 25% fatigue allowance.

The weight has to be picked up every minute. If the frequency were once every 5 minutes, the 50 pounds would be divided by 5.

$$5\% + (\tfrac{4}{5} \times 5) = 9.0\%$$

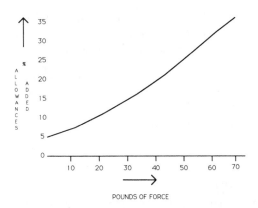

FATIGUE CURVE

% ALLOWANCES / FORCE IN POUNDS

FIGURE 9-26 Fatigue allowance curve: % allowances per pound of force

1. 5% IS MINIMUN FATIGUE ALLOWANCE
2. 5% OF INCREASE FATIGUE ALLOWANCES PER 10 POUNDS
    OF INCREASE FORCE OVER 10 POUNDS.
3. FORCE IS WEIGHT OF PART IF LIFTED.

The basic fatigue allowance is still 5%. When only lifting one fifty pound object every five minutes only one fifth of the excess weight is considered. 40 pounds is 4 units of weight, so:

$$\frac{4 \text{ units} \times 5\%}{5} = 4\% \text{ excess fatigue allowance}$$

5% Basic + 4% Excess equals 9% fatigue allowance

The duration of breaks must now be calculated. The normal 5% fatigue allowance is commonly interpreted as two 12-minute breaks, one in mid-morning and one in mid-afternoon, or a combination of the two, adding up to 24 minutes. Five percent of the 480 minutes in an 8-hour day is 24 minutes.

Seventeen percent allowances would equal 82 minutes per day. How will this 82 minutes be split for frequency and duration of breaks? I suggest that 11 minutes be given every hour except the hour before lunch. Seven 11-minute breaks equals 77 minutes, plus a 5-minute clean-up at the end of the shift. Note that a heavy job such as the one we are discussing here will tire out the employee faster than light or mental work, and the increased breaks are not only justified but will increase production. Breaks from work allow employees to recuperate, so when they return to work their production rate is higher than it would have been without a work break. The break more than pays for itself.

1. 5% is minimum fatigue allowance.
2. 5% increased fatigue allowances per 10 pounds of increased force over 10 pounds.
3. Force is weight of part if lifted.

Delay Allowances Are called unavoidable because they are out of the operator's control. Something happens to prevent the operator from working. The reason must be known and the cost accounted for to develop the cost justification.
Examples of unavoidable delays include

1. Waiting for instructions or assignments
2. Waiting for material or material-handling equipment
3. Machine breakdown or maintenance
4. Instructing others (training new employees)
5. Attending meetings, if authorized
6. Waiting for set-up. Operators should be encouraged to set-up their own machines. A set-up is complete when quality control approves.
7. Injury or assisting with first aid
8. Union work
9. Reworking quality problems (not operator's fault)
10. Nonstandard work—wrong machine or other problem
11. Sharpening tools
12. New jobs that have not been time-studied yet.

The operator's performance must not be penalized for problems out of his or her control. (Delays that are controlled by the operator are called personal time and are not considered here.)
Three methods are available to account for and to control unavoidable delays:

1. Add delay allowances to the standard.
2. Time-study them and add to the time standard.
3. Charge the time to an indirect charge (see chapter 12 for a more detailed discussion of this).

The goal of time study is to eliminate delay allowances. This is best done by time-studying the delay and adding that time to the time standard. However, some delays are so complicated that negotiating an allowance with the operator will save time and money for the company. For example, how much time do you spend a day cleaning the machine? The operator will always say, "Well, it depends," and the technologist must ask something like

What is the longest time?

What is the shortest time?

Do you think 15 minutes is a good average?

If the operator agrees that 15 minutes per day is a good figure, the technologist will calculate a delay allowance as follows:

$$\frac{15 \text{ minutes clean-up}}{480 \text{ minutes/shift}} = 3\%.$$

A 3% allowance will be added to the personal allowance of 5% plus a fatigue allowance of 5% to produce a 13% total allowance.

Generally, unavoidable delays can be eliminated or anticipated. Time standards in the form of standard data can be established and added to the time study to compensate the operator. An unavoidable delay is a foreign element, discussed later in this chapter. Those unavoidable delays that cannot be anticipated will require the operator to charge their time to an indirect account—meeting, injury, machine breakdown, and rework are examples. Supervisors will be required to approve all indirect charges, and the time should be more than 6 minutes to be statistically significant. The employee must not be penalized for management's lack of planning, but the supervisor must be given as much advance notice as possible. Reassignment may be in order.

One last warning about delay allowances: Don't put anything in the time standard that you cannot live with. It is difficult to get it out of the standard once it is included. Most companies have eliminated delay allowances but have allowed operators to punch out for anything not covered by the time standard.

Personal fatigue and delay allowances are added together, and the total allowance is added to the normal time:

Normal time + allowance = standard time.

METHODS OF APPLYING ALLOWANCES

Allowances are added in four different ways. The forms in this text use just one of these methods, but there are good reasons for using the other methods. Each company has its own time study form and procedure. The form tells you which method of applying allowances to use. The four methods are presented here in order of ease of application.

Method 1: 18.5 hrs per 1,000

This method is the simplest method of all and reduces the mathematic steps. It is also based on a constant allowance—in this case, 10%.

If a job takes 1.000 minutes normal time, how many pieces per hour could be produced? At the rate of one per minute, 60 could be produced per hour, but we want to be practical and add 10% allowances. Ten percent of 60 is 6, so 54 pieces per hour would be an appropriate time standard. How many hours would it take to produce 1,000 units at the rate of 54 per hour? One thousand divided by 54 equals 18.5 hours per 1,000—the name of this method. Three numbers are required to communicate a time standard:

$$\text{Decimal minute} = 1.000$$
$$\text{Hours per } 1,000 = 18.5$$
$$\text{Pieces per hour} = 54.$$

All time standards start with a decimal minute, so if our next standard is .5 minutes, the hours per 1,000 equals $.5 \times 18.5 = 9.25$ hours per 1,000, and the pieces per hour is $1/x$ or 108 pieces per hour. Try these examples:

Normal Minutes	Hours/1,000 18.5	1/x Pieces per Hour
.250	4.625	216
.333		
.750		
1.459		
2.015		

Notice that no calculations are made adding allowances; it is all in the 18.5.
What would the hours per 1,000 be with 15% allowances?

Method 2: Constant Allowance Added to Total Normal Time

This is the method used in this text and the most common technique used in industry.
Each department or plant has only one allowance rate. The average allowance is between 10 and 15%. An explanation of what makes up the allowance must be put forth, such as

$$\text{Personal time} = 24 \text{ minutes}$$
$$\text{2 breaks @ 10 min.} = 20 \text{ minutes}$$
$$\text{Clean-up time} = \underline{4 \text{ minutes}}$$
$$\text{Total allowances} = 48 \text{ minutes}$$

$$\frac{48 \text{ minutes}}{480 - 48 \text{ minutes}} = 11\%.$$

Eleven percent is added to normal time to get standard time, or 111% times normal times equals standard time.

$$1.000 + .11 = 1.110 \text{ minutes}$$
$$1.000 \times 111\% = 1.110 \text{ minutes}$$

The time study form will tell you which calculation to make.

Method 3: Elemental Allowances Technique

The theory behind this technique is that each element of a job could have different allowances, as in the following example:

Element Description	Unit Normal Time	Allowance	Standard Time
1. Load machine	.250	15%	.288
2. Machine time	.400	5%	.420
3. Unload machine	.175	10%	.193

Note that each element allowance is different. Element 1 is operator controlled, and a heavy part is involved. Therefore, more allowances were involved. Element 2 is a machine element, and the operator just stands there—no fatigue was given. Element 3 is a normal 5% fatigue, plus 5% personal allowance.

    The obvious advantage of this method is improved elemental time standards. The disadvantage is the increased math effort required. The time study form would have to be redesigned to accommodate this method and, as with all allowances, the form would show you which technique to use.

Method 4: The PF&D Elemental Allowance Technique

As in Method 3, the allowance is placed on each element, and this method shows everyone exactly how the allowance was developed. This technique is the most complete of all techniques.

Example:

Element Description	Unit Normal Time	Allowances % P	F	D	Total	Standard Time
1. Load machine	.250	5	10	0	15	.288
2. Run machine	.400	0	0	5	5	.420
3. Unload machine	.175	5	5	0	10	.193

This allowance technique takes a lot of time and effort. It is very descriptive, but the cost is too high for most companies.

    Allowances are an important part of the time standard, and properly established allowances will assist in the continued improvement of the quality of work life. If a job has undesirable aspects that do not reflect on the individual cycle, the allowances must reflect this undesirability. In that way, the money exists to justify a needed change. A plant-wide base rate of 10% is still very desirable, but additional allowance can be added as needed. The forms used in this text allow for a range of allowances.

    Allowances are negotiated, while foreign elements are time standards for the same thing. Time study is best, but allowances are quicker to set.

FOREIGN ELEMENTS

Foreign elements are any elements of work not planned for by the time study technologist. Foreign elements may be absolutely necessary, but they don't occur every cycle and they may not be known when the time standard is set. There are two basic types of foreign elements:

1. Productive foreign elements
2. Nonproductive foreign elements.

Productive Foreign Elements

Productive foreign elements are necessary jobs that must be performed or the operation halts. Some examples are

1. Cleaning the chips or slugs out of a machine
2. Loading parts into a feeder
3. Moving finished material out and new material into the work station
4. Changing tools
5. Loading welding rod coil into welder.

These examples can be considered unavoidable delays, and we eliminate them by time-studying them.

Figure 9-27 shows an example of a productive foreign element.

Extend this study. Subtract the previous time from each time as always, but be careful. When did we clean the machine? Between 1.50 and 2.40, or .90 minutes cleaning time. When did we inspect a part? On the fourth cycle between 3.57 and 3.89, or .32 minutes inspecting time. Once the element cycle time is calculated, the frequency must be determined. How many parts can we run before the machine needs cleaning? This is discussed with the operator. How many parts are run before we check a part? Quality control procedures tell us this.

FIGURE 9-27

| | Descriptions | | | | |
Element Description	1	2	3	4	5
1. Load machine	16	83	150	3.17	4.15
2. Machine time	56	1.23	2.90	3.57	4.55
3. Unload	66	1.33	3.01	3.99	4.64
4. Clean machine			2.40		
5. Inspect parts				3.89	

$$\text{Clean machine} = \frac{.90 \text{ min.}}{300 \text{ parts}} = .003 \text{ min./part}$$

$$\text{Inspect} = \frac{.32 \text{ min.}}{10 \text{ parts}} = .032 \text{ min./part}$$

These times are added to the total normal time just like any other element of work.

One time study may not give the technologist enough information to set a good quality time standard for these foreign elements, but with enough time studies, eventually these foreign element times will start making sense.

Nonproductive Foreign Elements

Nonproductive foreign elements are eliminated from the time study. A nonproductive foreign element is a goof that should not be a part of the operation, such as

1. Dropping a part or fumbling

2. Stopping to talk to the time study technician

3. Tying a shoe

4. Time study technologist did not get a good reading.

The real question is, "Should this be a part of the standard?" The continuous time study requires the technician to record everything that happens during the study, but everything does not have to be included in the time standard. When something unusual happens during a time study, the technician places an asterisk (*) next to the ending point and describes what happened in the foreign element block ㉗. Once the time study is extended, the foreign element being discarded is circled. By circling the bad element time, it is highlighted but not obliterated.

One more point regarding foreign elements: If the operator performs a productive foreign element but the time study technician doesn't realize it in time to record the ending point of the previous element, it is not a problem. The technician enters the next normal ending point, marks the reading with an asterisk, and discards that reading because it contains work time and the foreign element time. Our objective is to set a good time standard for a job, so throw out any element of work that detracts from this goal. An explanation in the foreign element block ㉗ is important.

LONG CYCLE TIME STUDY

The long cycle time study worksheet is used for

1. Long cycle time—15 minutes and longer

2. Inconsistent element sequence

3. Eight-hour performance studies.

FRED MEYERS & ASSOCIATES LONG CYCLE TIME STUDY WORK SHEET

PART NO. Quart Line
OPERATION NO. Line
DATE/TIME 10/10/xx
BY I.E. Meyers

OPERATION DESCRIPTION: 4 people (loader, Machine, Cartons, Unloader) Automatic Quart Line Cann
MACHINE; TOOLS. JIGS: #1--300 CANS/Minute
MATERIAL: Motor oil any weight

ELEMENT #	ELEMENT DESCRIPTION Started 7:00 AM Ended 3:30 PM	ENDING WATCH READING	ELEMENT TIME	% R	NORM. TIME
1	Shift Start up--No production	7:05	5.0	100	5.0
2	Run	7:06	1.0	100	1.0
3	Stopped--operator forgot something	7:07½	1½	0	----
4	Run	7:14	6½	100	6.5
5	No Lids	7:16½	2½	0	----
6	Run	7:19	2½	100	2.5
7	Check Temperature	7:20½	1½	70	1.05
8	Run	24	3½	100	3.5
9	Box Jammed in Former	25	1	110	1.1
10	Run	28¼	3¼	100	3.25
11	Bad Can	29	3/4	120	.9
12	Palletizer Jam-Bad Pallet	31	2	110	2.2
13	Run	33	2	100	2
14	Bad Box in Former	7:33½	½	130	.65
15	Run	7:41	7½	100	7.5
16	Bad Box in Former	7:41½	½	140	.7
17	Run	7:51	9½	100	9.5
18	Bad Box in Former	7:54	3	120	3.6
19	Run	7:56	2	100	2.0
20	No Lids in Machine	8:00½	4½	0	----

FIGURE 9-28 **Long cycle time study: Page one of eight.**

On long cycle jobs, many foreign elements tend to enter the study, and the sequence is not the same every time. These problems tend to create poor results from the continuous and snapback form.

Figures 9-28 and 9-29 show an example of an 8-hour time study. The data was collected on eight pages of the long cycle time study worksheet and summarized on the graph. This example study was taken on a quart oil canning line. The purpose of the study was to determine and eliminate the reasons for stopping the line. In one hour, 24,000 quart cans could be filled and packed—24 per case. One thousand cases per hour was the potential of the line, but 3,500 cases per 8-hour shift was the average output. The 8-hour time study showed seventy-two work stoppages in one 8-hour shift. Each work stoppage was recorded, timed, and the reason for stoppage recorded. Once the problems were identified and costed, economical solutions could be found and implemented. The resulting improvements are immediate and financially significant.

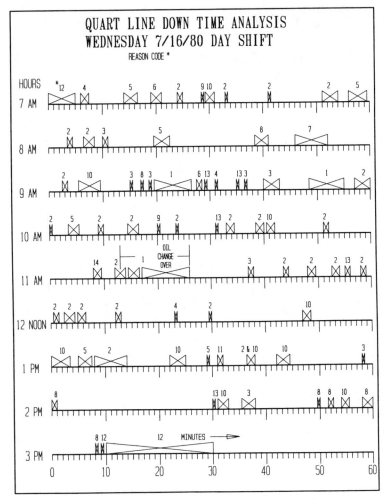

FIGURE 9-29 Graphical analysis of 8-hour time study.

STEP-BY-STEP INSTRUCTIONS FOR PREPARING THE LONG CYCLE TIME STUDY WORKSHEET

See Figure 9-30 for an example of the worksheet.

① Part no.: The part number identifies the part being studied.

② Operation no.: The operation number identifies the specific operation on the part being studied.

FRED MEYERS & ASSOCIATES			LONG CYCLE TIME STUDY WORK SHEET			

PART NO.	1	OPERATION DESCRIPTION:				
OPERATION NO.	2		5			
DATE/TIME	3	MACHINE; TOOLS, JIGS:	6			
BY I.E.	4	MATERIAL:	7			

ELEMENT #	ELEMENT DESCRIPTION	ENDING WATCH READING	ELEMENT TIME	% R	NORM. TIME
8	9	10	11	12	13
				14
				15
				16

FIGURE 9-30 Long cycle time study worksheet: The step-by-step form.

③ Date/time: The date of the study and the time of day the study was started.

④ By I.E.: The name of the time study technician doing the study goes here.

⑤ Operation description: A detailed description of the operation being performed is needed to communicate to future generations of operators, managers, and technicians what was being done when the standard was set.

⑥ Machine: Tools/jigs. Any machine, tool, or jig.

⑦ Material: The material specification may have an effect on machine speeds and feeds, so this information is important. Blueprints of parts are desirable and could be attached to the study.

⑧ Element #: The element number is just a sequential number and is used for reference.

⑨ Elemental description: The elemental description in long cycle time study may be lengthy, but it is important to know what is included in the time standard. Be as descriptive as possible and include sizes, number of parts, weight, any unit of measure that may be the reason for time to vary. A good rule is, "You can't have too much information."

⑩ Ending watch reading: As in all time study, the ending time is the only time to be recorded. The continuous time study technique is used.

⑪ Element time: Subtract the previous time from each watch reading to calculate element time. This is the time it took to perform this element of work.

⑫ % R: Percent rating is the technologist's opinion of the speed and/or tempo of the operator. The speed of the operator can change the time required significantly. We must always rate the operator on each and every element of work.

⑬ Normal time:

$$\frac{\text{Element time} \times \% \text{ R}}{100} = \text{normal time}$$

The time it should take an average person working at a comfortable pace.

⑭ Total normal time: Even though this block on the form is not labeled, this is where it should go. The reason that total normal time does not appear at the bottom of the form is that many pages of long cycle time study worksheets are involved, and the last page is the only one that needs the total normal time block. This is the same location and math used on the previous time study & PTSS forms.

⑮ Standard time: Normal time plus allowances equals standard time.

⑯ Hours per unit: Standard time divided by 60 minutes equals hours per unit.

⑰ Pieces per hour: Pieces per hour is the $1/x$ of hours per unit, or divide the hours per unit into 1.

TIME STUDY PRACTICES AND EMPLOYEE RELATIONS

Time study practices and employee relations can be summarized in a list of practices and attitudes that have been developed over the years to promote and improve the time study technician's image and improve results.

1. Always subtract the previous watch reading from each watch reading using a red pen. The elemental time is more important than the watch reading, and extending the study in red will highlight the elemental time. Only the subtraction needs to be in red.

2. Always stand up while taking a time study. Sitting or slouching presents a lazy, nonproductive attitude. The time study technician is a leader in productivity improvement, and presenting the right attitude is important.

3. Talk to the operator. The operator is the expert on the job you are time studying and therefore is the source of much information you need. Asking questions shows respect, and respect is what we all want. There is no room in our profession for time study technicians who do not talk with people. They may be shy, but employees see them as being stuck up.

4. Be positive about your time standards. If you don't believe in your time standards, who will? If you allow doubt about your time standards, they will not be as useful. You need to take every opportunity to sell the accuracy of your standards and defend them quickly. If needed, restudy the challenged job immediately. There can be no excuse for not achieving the time standard.

5. Get the supervisor's permission to enter his or her area. The supervisor's area is like a home, and you wouldn't enter someone's home without permission. The employee works for the supervisor, not the time study technician. If anything needs changing, have the supervisor give instructions.

6. Try to put the operator at ease. Tell the operator what you are going to do and why. The operator may have a bad attitude about time study, and it is your job to sell yourself as an honest person.

7. *Be* an honest person. The objective of a good time study technician is to set fair and equitable time standards. Nothing should interfere with this goal. Tell operators that you are honest and wouldn't knowingly do anything to hurt them.

8. Be a friendly and happy person. The negative attitude of an employee cannot affect your attitude. Make it a challenge to win over even the most obnoxious employee. This person was probably hurt by a time study in the past, but you can correct those errors.

9. Stand in a position where you can see what the operator is doing, but out of the way of moving machines or flying parts. Never hide while doing a time study. This is not professional.

10. Never change a time standard without good reason, and if a change is made

a. It must be over 5%;

b. It must be communicated to the employees—The reason for the change, and the amount of change.

This is an ever-growing list, and if you think of something that should be here, please share it.

QUESTIONS

1. Stopwatch time study can only be learned by doing. Identify and time-study two production operations using everything learned in this chapter.

2. Why is stopwatch time study the most popular method of setting time standards?

3. Why is time study a good entry-level job for an industrial technologist?

4. What are the six different stopwatches, and what are the benefits of each? Which one will we use and why?

5. What are the uses of a video camera in motion and time study?

6. What are the ten steps of the time study procedure?

7. Who should we time-study, and who should we not time-study?

8. What should the technologist check before starting a time study and why?

9. What are the eight principles of elemental breakdown?

10. What are the four reasons for breaking down a job into elements?

11. How is the frequency column ㉑ used? (Figure 9-13)

12. Determine the number of cycles needed for every element of your time studies in Question 1.

13. What are rating, leveling, and normalizing?

14. What are allowances?

15. What do we mean by "check for logic", and why is it important to do so?

16. What are the four factors of rating, and how important is each?

17. What are some of the standards used to define 100% performance?

18. Check your performance calculations on page 146 with Figure 9-22.

19. What is the rater trainer form?

20. What are the three categories of allowances? Give an example of each, and a typical percent allowance for each.

21. How does excessive weight affect the allowance?

22. How do we apply allowances?

23. What is a foreign element?

24. How do we handle foreign elements?

25. Time-study two projects. Show everything that is required in a proper time study. Use the continuous time study method.

CHAPTER 10

Standard Data

DEFINITION

Standard data is a catalog of elemental time standards developed from a database collected over years of motion and time study. The catalog of time standards is organized by machine names or numbers and job descriptions. When a new part is designed and the fabrication steps have been identified, the industrial technologist looks up the machine in the catalog. That machine page tells the technologist what causes the time to vary, so the technologist takes measurements from the blueprint of the new part and finds the time for this new job.

Each job may have several elements, so several elemental times will be developed for each job. When time studies were done, the time study technologist broke down the job into elements. The main reason for doing this, as stated in chapter 8, is to develop standard data. Different elements have different reasons for time to vary. Some elements are constant; some are variable. Some elements are machine controlled and some are operator controlled. Therefore, standard data times will be more accurate when broken down into elements.

Standard data times are for full elements, whereas PTSS consists of time data for smaller divisions of work—basic motions. It could take up to 30 minutes to develop an elemental time using PTSS, while the same standard could be set in one minute using standard data. PTSS and time study are used to develop standard data.

ADVANTAGES

The advantages of standard data time standards include the following:

1. Time standards are easier and more quickly set than stopwatch or PTSS techniques.

2. Time standards are more accurate.

3. Time standards are more consistent.

4. Time standards can be set before production starts.

5. Time standards for short duration jobs can be economically set.

6. Standard data reduces the need for time study.

7. Time standards are easier to explain and to adjust if needed.

8. The cost of time standard application is greatly reduced.

Standard data time standards are more accurate and consistent than other time standard techniques because individual differences between jobs are smoothed out in curves, formulas, or graphs.

Adjustments can be made quickly, if needed. A graph can be raised or lowered a percentage and change every job on that machine. A standard data element may be tight or loose, but all jobs will be tight or loose and can be corrected by a minor change.

METHODS OF COMMUNICATING STANDARD DATA TIME STANDARDS

Time standards can be presented in many forms, and this is the main subject of this chapter. The development of standard data is the ultimate job in motion and time study, and standard data development must be the goal of every Industrial Engineering department. Standard data development is much like detective work. The job is to find what causes time to vary, and the better the job the technologist did in breaking down the job into elements during the time study phase, the easier the standard data development will be.

The following methods of communicating standard data are discussed in this chapter:

1. Graphs

2. Tables

3. Formulas

4. Worksheets

5. Machine feeds and speeds.

Graphs

An example of graphs and a worksheet is shown in chapter 4, Figure 4-5. Please review this example.

The time study shown in Figure 10-1 is a classroom example of how a standard

FRED MEYERS & ASSOCIATES   TIME STUDY WORKSHEET   ☐ SNAP BACK ☒ CONTINUOUS

OPERATION DESCRIPTION: Walking (w) in paces (p) & Dealing Cards (c)

PART NUMBER 0052	OPERATION NO. 05	DRAWING NO. 0052-005	MACHINE NAME None	MACHINE NUMBER None	☒ QUALITY OK ?
OPERATOR NAME Fred	MONTHS ON JOB 60	DEPARTMENT Technology	TOOL NUMBER None	FEEDS & SPEEDS.	☒ SAFETY CHECKED ?
PART DESCRIPTION: Standard play cards		MATERIAL SPECIFICATIONS: Plastic coated		MACHINE CYCLE NONE / TIME	☒ SETUP PROPER ? NOTES:

ELEMENT #	ELEMENT DESCRIPTION	R/E	1	2	3	4	5	6	7	8	9	10	TOTAL CYCLES	AVERAGE TIME	%R	NORMAL TIME	FREQUENCY	UNIT NORMAL TIME	RANGE	R/X	HIGHEST
1	W5P	R	05	64	19	99	54	12	89	47	5.03	60	.41								
		E	.05	.04	.04	.05	04	05	25	05	04	05	9	.046	105	048	1	.048	.01	21	7
2	C15C	R	18	77	33	2.13	68	26	4.03	61	17	74	1.38								
		E	.13	.13	14	.14	.14	.14	.14	.14	.14	.14	10	.138	115	159	1	.159	.02	14	5
3	W10P	R	27	87	41	22	78	34	12	70	26	82	.89								
		E	09	.10	.08	09	.10	08	.09	.09	09	.08	10	.089	105	093	1	.093	.02	22	5
4	C10C	R	38	96	74	32	88	44	21	80	36	93	.90								
		E	.11	.09	.33	10	.10	10	.09	10	.10	.11	9	.100	115	115	1	.115	.02	20	0
5	W15P	R	52	1.09	89	46	302	59	35	93	50	6.07	1.40								
		E	.14	.13	.15	.14	.14	.15	.14	.13	.14	.14	10	.140	100	140	1	.140	.02	14	3
6	C5C	R	58	15	94	50	7	64	42	99	55	14	.58								
		E	06	.06	.05	06	.05	05	07	.06	05	.07	10	.058	120	070	1	.070	.02	28	6

FOREIGN ELEMENTS:
*Stop to talk
*Dropped deck

NOTES: This time study to be used for standard data.

ENGINEER: Meyers   DATE: 2 / 8 / xx
APPROVED BY: DATE: 2 / 9 / xx

R/X	CYCLES
.1	1
.2	14
.3	15
.4	27
.5	42
.6	61
.7	83
.8	108
.9	138
1.0	169

TOTAL NORMAL MIN. .625
ALLOWANCE + 11 % .069
STANDARD MINUTES .694
HOURS PER UNIT 0 11 57
UNITS PER HOUR 86

ON BACK
WORK STATION LAYOUT
PRODUCT SKETCH

FIGURE 10-1 Standard data time study: Three walking elements and three of counting cards.

data graph is created. The first step is to time-study (or PTSS) enough different jobs at a work station to determine what causes time to vary. This time study has three elements of walking (at different distances) and three elements of counting cards (of different amounts). Figures 10-2 and 10-3 show how the three data points look on a graph. Notice the perfect straight line. This is not normal. Usually, one point or more includes a small error, and for this reason many data points or element time studies should be made.

Once the graphs are produced, the next time a time standard is needed for a new walking or counting cards element, the technologist can go to the graph, look up the number of paces or cards, move up the graph to the line, and then across horizontally to the time. That is the normal time for the new job. Never again will we have to time-study walking or counting cards unless the new job exceeds the limits of the graph.

Tables

Whatever can be communicated with graphs can be converted to tables. When untrained people are required to use standard data, the table is the method of choice because it is

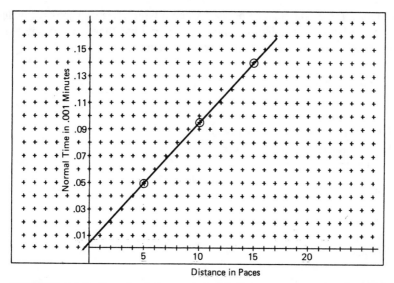

FIGURE 10-2 Standard data graph: Walking in paces, developed from Figure 10-1.

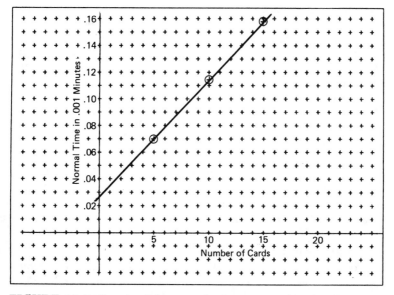

FIGURE 10-3 Standard data graph: Counting cards, developed from Figure 10-1.

easy to use. Tables should also be used when a large variety of jobs are included under one heading but have nothing in common. Material handling is an example of this. The following is an example of material handling standard data.

STANDARD DATA

Set-Up and Material Handling
Time in .001 Normal Minutes
The Following Data Are Only a Few of the Thousands Available:

1. Punch in and out of job, get blueprint, process sheet, reset counter, receive instructions	2.700
2. Plus walk to and from crib, supervisor, and time clock	.004/foot
3. Adjust adjustable gauge	2.340
4. Set up punch press	
15 min.	
30 min.	
60 min.	
5. Pick up and aside hand tool	
Small .060	
Large .078	
6. Loosen or tighten nuts or bolts	
With allen wrench	.084
With open or box wrench 1/2 −	.132
1/2 +	.192
Finger tighten	.078
7. Pick up and hand assemble nut or bolt	
From set-up table to machine	.480
From machine bed	.258
8. Remove and aside nut and bolt	
To set-up table	.318
To machine bed	.204
9. Pick up or aside punches, dies, gauges, etc.	
To or from set-up table	.192
To or from rack on machine	.234
To or from storage rack	1.320

A table of card counting and walking in paces is included in the next section ("Formulas") because the three elemental times are needed to develop formulas. The interchangeability of graphs, tables, and formulas should be evident. Any of the techniques can be used, but one technique will be better for one set of circumstances than another technique.

Tables are used in the automobile dealership service department for estimating work on your car. This time standards book is called the *flat-rate manual. Means Construction Guide* is another catalog of facts used by the construction industry. Look up the parts listed on the building blueprints, and the *Means Construction Guide* will tell you how many hours per unit each part takes. Cost estimating for construction is much

STANDARD DATA
MATERIAL HANDLING
PRODUCTIVE FOREIGN ELEMENTS

10. Move tub load of material out of a work station, move empty to other side, move new tub of material in (fork truck, hand truck, or hand jack)	1.000
11. Move box of parts into a work station by hand	.500
12. Pack parts in box	
1 at a time—large part	.150
2 at a time—medium parts	.050/part
Per additional part from same tub	.010/part
13. Form carton and	.150
Tape—manual	.100
Staple—4 staples	.050
Stitch—3 places	.075
14. Pick up loaded pallet (lifting time only)	
Manual—mechanical	.126
Manual—hydraulic	.402
Electric	.162
15. Position truck to pick up pallets	
Manual	.258
Electric	.090
16. Move load	
Manual	.003/foot
Electric	.004/foot
17. Set down loaded skid	
Manual—mechanical	.084
Manual—hydraulic	.192
Electric	.066

smarter and more accurate because of this catalog of time standards. Every industrial plant should be building its flat-rate manual.

Formulas

Many types of formulas are used in standard data. Straight-line, curvilinear, and special formulas are examples, but this chapter concentrates on straight-line. Formulas are another way of communicating graphical information. The formula for a straight line is as follows:

$$y = a + bX$$

y = vertical axis and measures time. In our work, normal time in thousandths of a minute (.001) is used.

a = the y intercept (where the line crosses the y axis). a is the time for getting started and finishing up. This time is required even if nothing else is done. In our dealing of cards, this is the amount of time required to pick up the deck and put it down.

b = slop of the line. This is the amount of time per unit of production. In the card-dealing example, this is the time per card.

X = the number of units of work. In card dealing, this would be the number of cards to be dealt. In the other example, it could be the size, weight, or number of paces. paces.

Let's take the card dealing example and ask how much time will it take to count out twenty-five cards:

$$x = 25, \ Y = ?, \ a = .025, \ b = .009$$
$$y = .025 + .009(25)$$
$$y = .250 \text{ minutes.}$$

The computer would be loaded with an element number for counting cards, and the a and b time standards would be assigned to that element. When a new standard is needed, the technologist would call up the element number and key in the quantity X, and the answer would appear immediately. Time standards for a whole new product could be set in a fraction of the time any other system would require.

Calculating the variables (a and b) is necessary. They could be estimated from a graph, but formulas would be much more accurate. Let's use the walking and counting cards examples. The graphs of walking and counting cards could be used to estimate a and b. (When we finish calculating them using formulas, please compare your estimate with the math answer.)

The formula for fitting a line to a set of data is called regression analysis. The x and y coordinates are used to calculate the variables a and b. The x and y data comes from time studies.

A table of walking and counting cards would look like this (again based on the time study):

Walking in Paces	Normal Time in Minutes	Count Cards	Normal Time in Minutes
x	y	x	y
5	.048	5	.070
10	.093	10	.115
15	.140	15	.159

The regression line equations to solve for a and b are

$$a = \frac{(\Sigma x^2)(\Sigma y) - \Sigma x(\Sigma xy)}{N(\Sigma x^2) - (\Sigma x)^2}$$

$$b = \frac{N(\Sigma xy) - \Sigma x(\Sigma y)}{N(\Sigma x^2) - (\Sigma x)^2}.$$

From the preceding table data, an extension of the data is needed:

WALKING

N	X	y	x^2	xy
1	5	.048	25	.24
2	10	.093	100	.93
3	15	.140	225	2.10
Σ	30	.281	350	3.27

$$a = \frac{350(.281) - 30(3.27)}{3(350) - (30)^2} = .002 \text{ minutes}$$

$$b = \frac{3(3.27) - 30(.281)}{3(350) - (30)^2} = .009 \text{ minutes.}$$

Therefore, the y intercept is .002 minutes and the unit time per pace is .009 minutes. The time required to walk 20 paces is $y = a + bx$. The a and b are now constants of .002 and .009 respectively, so $y = .002 + .009(20)$, or .182 minutes to walk twenty paces.

COUNTING CARDS

N	x	y	x^2	xy
1	5	.070	25	.350
2	10	.115	100	1.150
3	15	.159	225	2.385
Σ	30	.344	350	3.885

$$a = \frac{350(.344) - 30(3.885)}{3(350) - 30^2} = \frac{120.4 - 116.55}{1050 - 900} = \frac{3.85}{150} = .026 \text{ minutes}$$

$$b = \frac{3(3.885) - 30(.344)}{3(350) - (30)^2} = \frac{11.655 - 10.32}{1050 - 900} = \frac{1.335}{150} = .009 \text{ minutes}$$

So the counting of cards takes .026 minutes to pick up and put down the deck of cards and .009 minutes per card. How much time would it take to count twenty-five cards?

$$y = a + bx; \ y = .026 + .009(25); \ y = .251 \text{ minutes to count 25 cards}$$

Worksheet

A worksheet is much like a big formula. The job has many elements of work, some constant and others variable. The worksheet has questions and blanks to be filled in about the variables, and pretyped standards about the constants. The end of the worksheet is just like the PTSS form or time study form, in that total normal time, allowances, hours per unit, and pieces per hour are calculated. The advantages of a worksheet are that

1. Any clerk can set standards.
2. Standards can be set before production starts.

FIGURE 10-4 Standard data worksheet
parts-bagging operation

Model Number **1660**	Set Name **GAS GRILL**
No. of Different Parts **15**	Date **2-16-xx**
No. of Individual Parts **105**	Name of Technician _____
No. of Chains ___	Instruction Book No. **1660-1200**
Bag Part Number **1660-1550**	**1660-1250**

Element # Operation Description	Normal Time in Minutes
1. Get Bag, Open Bag, Place on Funnel	.052
2. Pack Parts	
A. # of Different Parts	**15** × .018 = _____
B. # of Individual Parts:	
Small Parts $\frac{1}{8}''$ or Less Thick	**50** × .011 = _____
Medium Parts $\frac{1}{8}''$ to $\frac{3}{4}''$ Thick	**30** × .008 = _____
Large Parts $\frac{3}{4}''$ and Over	**25** × .005 = _____
C. # Chains	___ × .075 = _____
D. # Pages or Booklets	**2** × .025 = _____
3. Remove Bag, Fold, Staple, and Aside	.110
4. Total Normal Time	_____
5. Plus 10% allowance (NT × 1.10)	_____
6. Hours per Bag = line 5 ÷ 60 minutes/hour	_____
7. Bags per Hour 1/x of 6	*_____

*Make the answer come out to 39 bags per hour.

The worksheet standard data system is easy to use and is a valuable technique.

An example of the use of a worksheet is the bag-packing operation of a local plant. (See Figure 10-4). This worksheet can be completed as a homework assignment. The bold numbers are from the parts list. This plant produces swing sets and gas grills. Over 100 different bags of parts are designed every year. One of their customers, Sears, requires the company to set its cost for the entire year before production starts. The price is included in the catalog, and changes in price are not allowed. How important is it to have a good time standard? Making a $1.00 error in cost estimating is just like including a dollar bill in the package. Making a $1.00 error in labor is like packing $3.00 in the box and shipping it.

Machine Speeds and Feeds

Some machines have constant cycle times (for example, mechanical punch presses). A technologist will hold down the cycle trip button and time ten cycles to get an average time. This time is recorded on a machine cycle time sheet for future reference. The cycle time could also be a constant part of a worksheet. It is used in PTSS. Constant cycle time is the easiest type of time standard to set and use.

Other machines have constant feed rates (for example, welding). A table provided by the manufacturer would tell the technician that a $\frac{1}{4}$-inch fillet weld-single pass will weld at the rate of 12 inches per minute. If 30 inches of weld are needed; 30 divided by 12 equals 2.5 minutes of weld time. Handling time and moving the welding whip from point to point must be determined by other means, machine time is just the machine cycle time.

The speeds and feeds of chip cutting machines, such as lathes, drills, and mills, depend on

1. Material being cut;
2. Type of tool being used.

Speeds and feeds for all of the aforementioned machines are scientifically determined to minimize cost, and the source of speeds and feeds information is the *Machinery's Handbook,* published by Industrial Press. (New York, N.Y.) Other sources for speeds and feeds information are tool and machinery manufacturers. Some manufacturers provide technologists with slide rules and plastic laminated speeds and feeds tables.

Speeds and feeds are the basis of machine run time. The time standard includes the machine run time and the load/unload time, which is controlled by the operator. A time study would be needed to determine the load/unload time, and after enough time studies have been made standard data tables could be developed.

Speed rates are specified in feet per minute. If a speed rate of 500 feet per minute is called for, a point on the diameter of the tool or part must travel at that speed. For example, if a 2-inch diameter bar was placed in a lathe, a spot on the diameter would need to be moved at a speed of 500 feet per minute. A 2-inch drill would also have to move a point on its outside diameter 500 feet per minute. A 2-inch diameter part or tool is a little over 6 inches around (πD), and the machine would have to turn this 6 inches (half foot) 1,000 times per minute to produce 500 feet per minute. This is the logic of the RPM (revolutions per minute). The formula is as follows:

$$*(1)\ \text{RPM} = \frac{\text{speed rate}}{\pi \text{D}}.$$

Speed rate is in feet per minute. Multiply by 12 for inches per minute.

Example: Two-inch part with a speed of 500 feet/min. Two inches comes from the blueprint, and 500 feet per minute comes from the *Machinery's Handbook.*

$$*(1)\ \text{RPM} = \frac{500 \times 12}{\pi(2)} = 955\ \text{RPM}$$

This is close to our logic example.

Feed rates are given in inches of advancement per revolution of the tool or part. A .002-inch feed rate means advance the tool two thousandths of an inch every time

*(1) is RPM

(2) is #rev required

(3) is time

the tool turns one revolution. In our answer to the foregoing RPM formula, that tool would move $.002 \times 955 = 1.91$ inches per minute. If we were to turn a part 2 inches back, or drill a hole 2 inches deep, about one minute would be required. Again, this is a logic approach, but for more accuracy and speed, the formula for number of revolutions is needed.

(2) Number of revolutions needed = # rev.

$$(2) \ \# \ \text{rev.} = \frac{\text{length or depth of cut}}{\text{feed rate}}$$

$$(2) \ \# \ \text{rev.} = \frac{2''}{.002} = 1,000 \ \text{revolutions}$$

The time required to turn 1,000 revolutions when the speed of the tool is 955 RPM is a little over one minute. The formula for this is

$$(3) \ \text{Time} = \frac{\# \ \text{rev.(2)}}{\text{RPM(1)}}$$

$$(3) \ \text{Time} = \frac{1,000 \ \text{rev.}}{955 \ \text{rev./min.}} = 1.047 \ \text{minutes.}$$

Let's look at some specific examples.

Lathe: How much time would be required to turn the part shown in Figure 10-5 on a lathe?

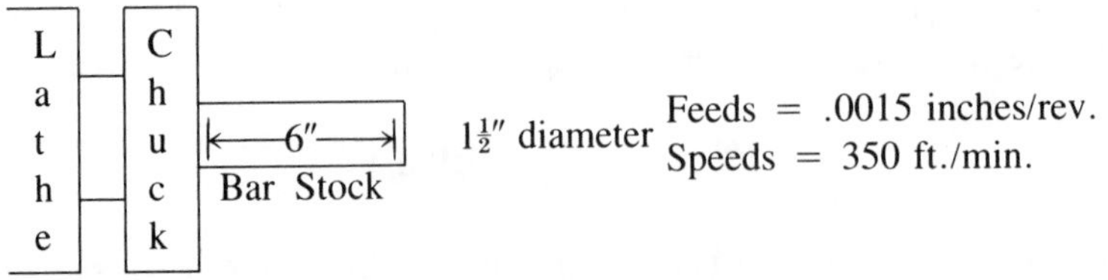

FIGURE 10-5

$$(1) \ \text{RPM} = \frac{350 \times 12}{\pi 1\frac{1}{2}} = 892 \ \text{RPM}$$

Let's be realistic. This machine can run 800 or 900 RPM, but not 892, so we will use 900 RPM.

$$(2) \ \# \ \text{rev.} = \frac{6''}{.0015} = 4,000 \ \text{rev.}$$

$$(3) \ \text{Time} = \frac{4,000}{900} = 4.444 \ \text{minutes}$$

Load/unload time is still required, but 4.444 minutes is the machine cutting time. Depth of cut may require two passes.

Drill: Drills are a little more complicated because of the effect of the drill tip on the length of cut. A drill tip is very close to .4 times the diameter of the drill. A $\frac{1}{2}$-inch drill would have a .2-inch drill tip ($.4 \times .5 = .2$). How much time will it take to drill a $\frac{3}{8}$-inch hole through a 2-inch part with a speed rate of 500 ft./min. and a .0025 feed rate? (See Figure 10-6.)

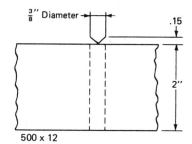

FIGURE 10-6

$$(1)\ \text{RPM} = \frac{500 \times 12}{\pi \frac{3}{8}} = 5{,}096\ \text{RPM}$$

Rounding off for actual machine capability = 5,000 RPM.

$$(2)\ \#\ \text{rev.} = \frac{2 + .15}{.0025} = 860\ \text{revolutions}$$

$$(3)\ \text{Time} = \frac{860}{5{,}000} = .172\ \text{minutes}$$

Again, the load/unload time must be determined by some other technique, because machine times are only cutting time, and the machine time is .172 minutes.

MILLS AND BROACHES

Mills are the most difficult of the three basic chip-producing machines because of

1. The overtravel
2. The number of teeth.

Regarding overtravel, the mill diameter can be quite large, and the diameter must be added to the length of cut if cutting through the part. One half of the diameter would account for cutting through the part, but a tail cut often leaves an uneven surface under the trailing half of the cutter, and if surface finish is important, the whole cutter must pass through the part.

A cutter could have one, four, eight, sixteen, or thirty-two teeth, and some cutters have sixty-four. Each tooth can take a feed rate of its own.

Example: How much time would it take to mill off the surface of an engine head made of cast iron with a speed rate of 225 and a feed rate of .0025? (See Figure 10-7.)

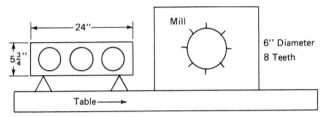

FIGURE 10-7

$$(1) \ \text{RPM} = \frac{225 \times 12}{\pi 6} = 143 \ \text{RPM, but 150 RPM Available}$$

$$(2) \ \# \ \text{rev.} = \frac{24 + 6}{.0025 \times 8} = 1{,}500$$

$$(3) \ \text{Time} = \frac{1{,}500}{150} = 10.000 \ \text{minutes}$$

Time standards resulting from standard data are in normal time. Allowances must be added afterward.

Standard data is the most efficient method of setting time standards. Every company should be setting standard data, and a good place to start is with a simple machine that is popular in that company. A toy company had a shrink wrap operation on every packout line. It had been setting standards on the shrink wrap job using PTSS and time study for years. The resulting standard data was developed in less than one day and reduced the standard-setting time from a 45-minute job to under 1 minute. This reduction in time-setting cost is typical. The industrial technologist has an unlimited number of problems to solve in any manufacturing plant, so don't make a career of time study. Standard data will put you out of the time study business.

QUESTIONS

1. Define standard data.

2. What are the advantages of standard data?

3. What are the five methods of communicating standard data? What is the advantage of each?

4. Calculate *a* and *b* for the following:

Size	Time
49″	.98
55″	1.10
75″	1.50
104″	1.80
115″	2.30

Note: One of these data points is in error. Plot a graph first and eliminate the error before calculating *a* and *b*.

5. Complete the worksheet example in Figure 10-4.

6. What is the time required to turn (lathe) a $\frac{1}{2}$-inch diameter part 3 inches back? The speeds and feeds are 250/ft./min. and .001 inches/rev.

7. What would be different if the part in Question 6 were a drill or mill?

CHAPTER 11

Work Sampling

Work sampling, TV sampling (Neilson ratings), political poll sampling (Gallup polls), and unemployment statistics are all random sampling, and as such are scientifically based on the same theories or laws of probability. Work sampling is the process of randomly observing people working to determine how they spend their time. Everyone who has ever worked with others has work sampled. The attitudes developed about fellow employees' work ethics or productivity are based on random observations. Conclusions are drawn that this person is a "work horse" or a "goof-off" because of random observations. Supervisors are work sampling their employees all the time. These informal work samples can be much more scientific and fair if done properly. The science and technique of work sampling is the subject of this chapter.

Work sampling is subdivided into three subtechniques. Each technique gets more complicated and must be learned in this sequence. Each technique is also a usable tool in itself.

1. Elemental ratio studies
2. Performance sampling studies
3. Time standard development studies

A time standard development study uses elemental ratio and performance sampling studies.

ELEMENTAL RATIO STUDIES

The primary task (what the operator does most of the time) performed by a person defines his or her job title, but many other activities (productive and nonproductive) take time, too. Each activity must be measured and compared to the total time. This is the element ratio. An elemental ratio study will determine the percentage of time each element of work takes.

Elemental Breakdown and Ratio Estimates

When starting an elemental ratio study, the elements of work must be listed, and the ratios must be estimated. A few observations are made before the study starts to estimate these ratios, but the estimates must be made to determine the total observations needed for a specific confidence level and accuracy. An example of a list of elements and the estimated elemental ratios follows:

Element No.	Elemental Description	Elemental Ratio
1	Load and unload	20%
2	Machine time	35
3	Set-up	15
4	Tool change	7
5	Inspection	5
6	Material handling	4
7	Away and idle	14
	Total	100

This study may be for a person, department, or a whole plant. It makes no difference to the study.

The smallest percentage (4% in this case) will determine the total number of observations required for a confidence level and accuracy, because larger percentages require fewer samples. Before discussing how many observations are required, confidence level and accuracy must be defined.

Confidence Level

The confidence level refers to how positive (confident) the technologist wants to be in the resulting ratios. At the beginning of a study, estimates are made based on very little information. There is little confidence in these ratios, but as data is collected, confidence builds. Every day the ratios are more consistent. The question at the beginning of the study is, How many observations do we need to take to be at a specific level of confidence? A 95% confidence level indicates our ratios are accurate 95% of the time. The remaining 5% will be off one way (over) or the other way (under), but not by much.

Accuracy

Accuracy measures the closeness of the ratio to the true ratio of an element. A ±5% accuracy indicates that the ratio is within 5% of the true element time. If a true ratio is 25%, the ±5% accuracy would allow the technologist's ratio to be ±1.25% above or below 25% (23.75 to 26.25%). A ratio of 10% ±5% equals 9.5 to 10.5%.

The goal of most motion and time study work is a 95% confidence level at ±5% accuracy. The technologist could produce 99% ±1%, but the cost would be prohibitive (more on this later in this chapter).

Sample

A sample is an observance of one operator one time. The observance of 100 operators one time each is exactly the same as observing one operator 100 times. Both produce 100 samples. Work sampling is observing enough employees enough times to collect the number of samples required to achieve the confidence and accuracy designed into the study. An observation of the operator must be made at the first moment of sight, and this observation must be made at random times—no preplanning. The science of sampling is based on the theory that a sample taken at random tends to have the same characteristics as the whole population. The work sampling of an operator, then, will be an indicator of how the operator spends his or her day. One sample won't do it, but many samples will.

Randomness

Randomness is a requirement of sampling. The exact time of an observation must be completely random (based on chance only), or the study's confidence and accuracy will be destroyed. The opposite of randomness is routine and predictability, and both destroy the study. Randomness can be developed many ways, discussed later in this chapter.

Sample Size: The Number of Observations

Sample size is determined by a combination of accuracy, confidence, and element percentage. The table in Figure 11-1 was developed from the following formula:

$$N = \frac{Z^2(1-P)}{(P)(A^2)}$$

N = number of observations needed

Z = number of standard deviations required for a specific confidence level (can be found in any statistics book).

Confidence Level	Z
99.5	3.25
99	2.575
95	1.960[a]
90	1.645
80	1.245
75	1.151

[a]95% confidence is used for most of time study, and the Z for 95% is 1.96.

FIGURE 11-1 Sample size table

How Many Samples Are Required?[a]

P	90% Confidence Z=1.645 Level of Accuracy			P*	95% Confidence Z=1.960 Level of Accuracy			P	99% Confidence Z=2.575 Level of Accuracy		
	1%	5%	10%		1%	5%	10%		1%	5%	10%
1	2,678,965	107,159	26,790	1	3,803,184	152,127	38,032	1	6,564,319	262,573	65,643
2	1,325,952	53,038	13,260	2	1,882,384	75,295	18,824	2	3,249,006	129,960	32,490
3	874,948	34,998	8,749	3	1,242,117	49,685	12,421	3	2,143,902	85,756	21,439
4	649,446	25,978	6,494	4	921,984	36,897	9,220	4	1,591,350	63,654	15,914
5	514,145	20,566	5,141	5	729,904	29,196	7,299	5	1,259,819	50,393	12,598
6	423,944	16,958	4,239	6	601,851	24,074	6,019	6	1,038,798	41,552	10,388
7	359,515	14,381	3,595	7	510,384	20,415	5,104	7	880,926	35,237	8,809
8	311,193	12,448	3,112	8	441,784	17,671	4,418	8	762,522	30,501	7,625
9	273,609	10,944	2,736	9	388,428	15,537	3,884	9	670,430	26,817	6,704
10	243,542	9,742	2,435	10	345,744	13,830	3,457	10	596,756	23,870	5,968
15	153,341	6,134	1,533	15	217,691	8,708	2,177	15	375,735	15,029	3,757
20	108,241	4,330	1,082	20	153,664	6,147	1,537	20	265,225	10,609	2,652
25	81,181	3,247	812	25	115,248	4,610	1,152	25	198,919	7,957	1,989
30	63,141	2,526	631	30	89,637	3,585	896	30	154,715	6,189	1,547
35	50,255	2,010	503	35	71,344	2,854	713	35	123,140	4,926	1,231
40	40,590	1,624	406	40	57,624	2,305	576	40	99,459	3,978	995
45	33,074	1,323	331	45	46,953	1,878	470	45	81,041	3,242	810
50	27,060	1,082	271	50	38,416	1,537	384	50	66,306	2,652	663
55	22,140	886	221	55	31,431	1,257	314	55	54,251	2,170	543
60	18,040	722	180	60	25,611	1,024	256	60	44,204	1,768	442
65	14,571	583	146	65	20,686	827	207	65	35,703	1,428	357
70	11,597	464	116	70	16,464	659	165	70	28,417	1,137	284
75	9,020	361	90	75	12,805	512	128	75	22,102	884	221
80	6,765	271	68	80	9,604	384	96	80	16,577	663	166
85	4,775	191	48	85	6,779	271	68	85	11,701	468	117
90	3,007	120	30	90	4,268	171	43	90	7,367	295	74
95	1,424	57	14	95	2,022	81	20	95	3,490	140	35
99	273	11	3	99	388	16	4	99	670	27	7

[a] Most work sampling studies use a 95% confidence level and a 5% accuracy. A sample is one observation.

*P = the percentage of the total that this element represents.

P = elemental percentage. A job may have several elements, but only the smallest element of an operation is used. In the example earlier in this chapter, material handling was 4% of the day, the smallest percentage of all the elements.

A = the accuracy desired. Most of the time, the study uses $\pm 5\%$ accuracy.

How many observations would we have to make in our example of 95% confidence with $\pm 5\%$ accuracy on a job that accounts for only 4% of the work day?

$$N = ? \quad Z = 1.96 \quad P = .04 \quad A = .05$$

$$N = \frac{(1.96)^2(1 - .04)}{.04(.05^2)} = 36,897 \text{ observations}$$

Check the table in Figure 11-1: 95% $\pm 5\%$ for a 4% job. Check the table for a 5% job (inspection): 29,196 observations. Only the smallest element percentage is used because all higher numbers will give fewer cycles.

The new technologist often asks why 95% $\pm 5\%$. The answer is the cost of quality. An uninformed manager may ask for 99% $\pm 1\%$ standards. From the sample size table in Figure 11-1, an elemental ratio *(P)* of 1% would call for 6,564,319 observations. At a rate of 800 observations per hour (a good rate for a plant-wide study), 8,205 hours of the technologist's time will be required. If the technologist is paid $20.00/hour, $164,100.00 would be the cost of the study. A 95% $\pm 5\%$ study on a 2% job would take 94 hours or $1880.00. The cost of quality in this case was nearly $162,220.00. Let's be more realistic and cost conscious.

Probability and Normal Distribution

Work sampling is based on the laws of probability. The normal distribution curve is used to describe the laws of probability: an example of a poorly trained operator (inconsistent) working on a job has the following time results (Figure 11-2). Also, please "Review 9-17 & 9-18".

Any one observation has the following chances:

Between .99 and 1.01 = 20% chance (20 of 82)

Between .96 and 1.04 = 61% chance (50 of 82)

Between .93 and 1.07 = 85% chance (70 of 82).

The chances of one observation being below .93 or above 1.07 are $7\frac{1}{2}\%$ each, but the more observations taken, the closer our estimated average will be to the real (true) average.

Work sampling takes all samples (observation), adds them together, divides by the number of samples taken, and the average time results. The more samples, the better the answer.

FIGURE 11-2 Work/sampling data collection sheet

Cycle Time in Minutes	No. Observations	Normal Distribution Curve
.87, .88, .89	|	
.90, .91, .92	⊬⊬	
.93, .94, .95	⊬⊬ ⊬⊬	
.96, .97, .98	⊬⊬ ⊬⊬ ⊬⊬	
.99, 1.00, 1.01	⊬⊬ ⊬⊬ ⊬⊬ ⊬⊬	
1.02, 1.03, 1.04	⊬⊬ ⊬⊬ ⊬⊬	
1.05, 1.06, 1.07	⊬⊬ ⊬⊬	
1.08, 1.09, 1.10	⊬⊬	
1.11, 1.12, 1.13	|	
	(Probability) 1 5 10 15 20	
	No. Observations	

The odds of the next cycle being over 1.11 minutes or less than .9 minutes are very low.

Step-by-Step Procedure for an Elemental Ratio Study

Step 1:   Establish the purpose and goal of the study. For example,

1. Develop a better work load balance among team members.
2. Increase the productivity of a fork truck.
3. Reduce idle time.
4. Provide better, more economical service.
5. Reduce cost.

Match these goals with the following examples.

Step 2:   Identify the subject.

Examples:

1. A quart canning line produces 1,000 cases of oil per hour using four people. Is this the correct manning?
2. A fork truck driver supports the assembly line. We want to add two machines adjacent. Can the driver support the two new machines?

3. A tool crib always has people waiting. Should we add a new attendant? Sample the number of people in line with one attendant and two attendants.

4. The maintenance department wants to set up a satellite area next to a high-use area to reduce walking time.

Step 3: Identify the elements.

Examples: Match these with the examples in Step 1.

1. Work, idle

2. Transport loaded, empty, idle

3. Number of people in line

4. Walking, working, idle

Step 4: Estimate the elemental ratio percentages from Steps 1, 2, and 3.

1. Worker 1 (75/25); worker 2 (60/40); worker 3 (70/30); worker 4 (80/20)

2. 40/30/30

3. 3 or 4

4. 25/50/25

Step 5: Determine the level of confidence and accuracy. Most industrial engineering is based on 95% ±5%, but there is good reason to go to 90% ±10% if the small elements are unimportant. We will use 95% ±5%.

Step 6: Determine the number of observations. Looking at Step 4, example 1, 20% is the smallest elemental ratio. From the 95% ±5% table (Figure 11-1), 6,147 observations will be needed. Toward the end of the study, we may find the elemental time to be only 15%; 2,561 more samples will be needed if the same level of accuracy and confidence is to be achieved.

Step 7: Schedule the observations.

The schedule of observation is the work sampling plan. In addition, randomness can be achieved through scheduling trip times. The following examples describe how to schedule.

Example 1: Check the machine department's efficiency. There are twenty people in the department. A trip through the department and back takes 5 minutes. Because the observer has other work, only twenty trips will be made per day. Four hundred samples will be collected per day in less than 2 hours. When will we take these twenty trips?

First, break up the day into 5-minute increments (96 per 8-hour shift). Second, using a random number generator:

1. A two-digit random number table,

2. A random number button on a calculator,

3. Draw numbers from a hat,

4. Telephone book numbers (last two digits only).

Select the first twenty random numbers between 1 and 96. Convert these back to minutes of the day.

Third, put the times in order. This is the schedule of starting times.

Example 2: Continuously study the job in example 1. Randomness is introduced by varying the starting point and/or route through the department.

Example 3: When studying one person or small group in the same area where the observer works: a random electronic beeper can be programed to sound off x number of times per hour. When the beeper sounds off, the observer would look around and record what the person or people are doing.

Step 8:   Talk with everyone involved.

The supervisor and employees should be involved with the purpose and goals of the study. If they are in on the planning and goal setting, they know the problem that is being studied and why. In addition, they may be required to assist in collecting production counts and keeping the observer informed of absence and reassignment.

The supervisor and employees should be informed that being caught doing nothing is no problem because 10% idleness, personal time, or non-productive delays is expected.

The time keeper or payroll department should be consulted to record the correct hours worked during the study. The observation period should coincide with a payroll period.

Step 9:   Collect data, make observations.

In Step 3, the elements were identified. The observation can be recorded to the right of the element. In Step 2, example 2, a fork truck was estimated as 40/30/30.

Work Sample Fork Truck	Tue., Jan. 28, XX		%
40% transport loaded	ꜩꜩ ꜩꜩ ꜩꜩ ꜩꜩ ꜩꜩ ꜩꜩ ꜩꜩ ꞁꞁ ꜩꜩ ꜩꜩ ꜩꜩ ꜩꜩ ꜩꜩ ꜩꜩ ꜩꜩ	72	37.5
30% transport empty	etc.	63	32.8
30% idle	etc.	57	29.7
	Total	192	100

Step 10: Summarize and state conclusions.

Step 1 asked what the purpose was, and Step 2 asked what the subject was. Now, it's time to answer those questions.

The fork truck study showed

The truck was loaded 37.5% of the time.

The truck was moving, but empty 32.8% of the time.

The truck was idle 30% of the time (10% is normal).

In terms of hours,

The truck was moving material 3 hours per shift

$$(.375 \times 8 \text{ hours/shift} = 3 \text{ hours}).$$

The truck was moving, but empty 2.62 hours per shift

$$(.328 \times 8 = 2.62 \text{ hours}).$$

The truck was idle 2.38 hours per day (.8 hour is normal).

In conclusion, this truck could do at least 1.5 hours more work. Methods improvement and driver instruction could reduce the driving-empty ratio of 32.8%.

PERFORMANCE SAMPLING STUDIES

Performance sampling requires rating the operator when observing him or her. Rating was a major subject of stopwatch time study, and that is exactly what must be done in performance sampling. The observance of an operator happens in a moment, and in that moment the observer must judge the speed and tempo of the operator.

Operator speed and tempo vary from person to person, and for an individual operator the speed and tempo can vary from minute to minute. For work sampling, performance sampling fine tunes the ratios, making them more accurate.

For example, in a bag-packing operation, five independent work stations are available. From an elemental ratio study, the operation is divided into four elements:

Operation Description	Observation (%)								
	60	70	80	90	100	110	120	130	140
1. Bag packing	₮ℋℒ	ℋℋℒ ℋℋℒ	ℋℋℒ ℋℋℒℋℋℒ ℋℋℒℋℋℒ ℋℋℒℋℋℒ	ℋℋℒℋℋℒ ℋℋℒℋℋℒ ℋℋℒℋℋℒ	ℋℋℒ ℋℋℒ	ℋℋℒ III	ℋℋℒ I	II	I
2. Load hoppers				II	ℋℋℒ	III	II	I	
3. Away	ℋℋℒ II								
4. Idle	ℋℋℒ ℋℋℒ	ℋℋℒ	ℋℋℒ						

Elements 1 and 2 are productive and can be rated by placing a tally mark under the proper % heading. Ten percent increments are normal, and most rating is between 70 and 130%. Elements 3 and 4 are nonproductive and therefore cannot be rated. The step-by-step procedure is exactly the same as for elemental ratio studies until Step 10, the last step.

The data for this bag-packing example is summarized in Figure 11-3.

As each day's work is summarized and added to the previous totals, the ratios get closer to the accuracy and confidence level being sought.

During the one-month study of these five bag-packing stations, 825 hours of labor were used.

How many hours were spent on each element of the job?

	%	Hours	Explanation
Bag packing	61.50	507.38	$(.615 \times 825 \text{ hours})$
Load hoppers	10.25	84.56	$(.1025 \times 825 \text{ hours})$
Away	5.75	47.44	$(.0575 \times 825 \text{ hours})$
Idle	16.40	135.30	$(.1640 \times 825 \text{ hours})$
	93.9	774.68	
Efficiency loss	6.1	50.32	
	100	825	

What if during this 825 hours of bag packing the operators packed 35,392 bags? A time study has been performed, and a time standard can be set.

$$\frac{\text{Hours used}}{\text{Bags packed}} = \frac{507.38 + 84.56 \text{ hours}}{35,392 \text{ bags}} = .01673 \text{ hours/bag}$$

$$+ 10\% \text{ allow.} \quad .00167$$
$$\text{Standard time} \quad .01840$$
$$\text{Bags/hr.} \quad 54$$

TIME STANDARD DEVELOPMENT STUDIES

Work sampling can be used to develop time standards accurately and quickly. Time standard development studies pull together all the techniques of work sampling, and it is the ultimate use of work sampling.

The step-by-step procedure is exactly the same as for the elemental ratio study and the sampling study. The additional information needed is units produced and allowances. The time standard development system starts after the other two are completed.

FIGURE 11-3

	Work Sampling				Bag Pack			Fri., Feb. 18, XX		
	60%	70%	80%	90%	100%	110%	120%	130%	140%	Total
A. Bag packing										
No. Observations	5	10	15	20	15	8	6	2	1	82
Leveled observ.[a]	3	7	12	18	15	8.8	7.2	2.6	1.4	75
B. Load hoppers										
No. observations	—	—	2	5	3	2	1			13
Leveled observ.			1.6	4.5	3	2.2	1.2			12.5
C. Away—nonproductive										7
D. Idle—nonproductive										20
Total observations:	$82 + 13 + 7 + 20 = 122$									
A. Bag-packing ratio	$75/122 = 61.5$									
B. Load hopper	$12.5/122 = 10.25\%$									
C. Away	$7/122 = 5.75\%$									
D. Idle	$20/122 = 16.4\%$									
% productive	$75 + 12.5/122 = 71.7$									
Average rating of bag packing	$75/82 = 91.5$									
Average rating of loading hoppers	$12.5/13 = 96.2\%$									

[a]Leveled observations are comparable to normal time (% × no. observations).

FIGURE 11-4

Elemental Number	Job Description	No. Observations[a]	Ratio %[b]	Hours[c]
1	Millwright	20,000	40	1,536
2	Welder	5,000	10	384
3	Electrician	2,500	5	192
4	Machine repair	8,500	17	653
5	Carpenter	2,000	4	154
6	Walking	4,000	8	307
7	Away	3,000	6	230
8	Idle	5,000	10	384
	Total	50,000	100	3,840

[a] Number of observations resulted from the observations over a one-month period.

[b] Ratio % is the number of observations of one element divided by the total observations (20,000/50,000 = 40%).

[c] Hours: Total hours for the one-month study were made available from payroll (3,840 hours).

Example: A maintenance department was work sampled with the goal of setting time standards to a level of 95% ±5% on 3% elemental ratio *(P)* jobs. The resulting data was collected and is summarized in Figure 11-4.

If 40% of the maintenance time is millwright, then the maintenance department spent 1,536 hours working on millwright jobs (40% × 3,840 = 1,536 hours). This is pure 100% work—no idle time.

This is where elemental ratio studies and performance sampling studies would leave us. Now the production counts and allowances would be added. Figure 11-5 shows how we extend the data from Figure 11-4 to create time standards.

① These element numbers refer to Figure 11-4 element numbers. Only the productive elements have time standards.

② Hours were calculated in Figure 11-4.

③ Work order counts were collected and given to the industrial technologist by the maintenance supervisor.

FIGURE 11-5

① Element No.	② Hours	③ Work Order	④ Hours per Work Order	⑤ Plus 15%	⑥ Work Orders per Hour
1	1,536	825	1.86	2.14	.47
2	384	475	.81	.93	1.08
3	192	150	1.28	1.47	.68
4	653	55	11.87	13.65	.07
5	154	30	5.13	5.90	.17

FIGURE 11-6 Example: Work sampling data collection sheet

Work Sampling Observations

	Time of Sampling					
	8:15	*9:10*	*9:50*	*10:10*	*11:30*	*Total*
Die Cut	/	/		/	//	
Roll Inner End	/	//	/	//	//	
Tack Fiberglass	/	(	//	/	/	
Roll	/	/	/	//	///	
Nylon	/	/	/	/	/	
Mat. Handling		//	///	//	//	
Pull Mandril	////	////	////	//	/	
Denylon	/	/	//	//	//	
Drill			/	/		
Saw			/	/		
C. Grind	///	//	/	//	/	
Silk Screen	////	////	////	///	//	
Spray & Clean	/	/		//	///	
Assemble	//// //	////	//// /	//// //	///	
Packout	//	////	//	/		
Total Productive	28	30	30	30	23	141
No Activity	//// /	//// ///	////	//// ///	//// //// ////	
Walking	//	///	////	//	///	
Talking	//	/	//	///		
Away	///	/	////	//	///	
Total Nonprod.	13	13	15	15	21	77
Total Observations	41	42	45	45	44	217
% Performance	68%	71	67%	67	52	65%
						5
				+5% Allow.		70%

④ Hours per work order are calculated by dividing hours ② by work orders ③.

⑤ Plus 15% allowance. A management decision placed 5% of the 8% walking ratio into the allowance with 10% personal and fatigue time, creating a 15% personal, fatigue, and delay allowance. 115% times the hours per work order ④ equals standard time ⑤.

⑥ Work orders per hour: Work orders per hour is the $1/x$ of standard hours ⑤; or divide standard hours ⑤ into the whole number 1. An entire plant can be work sampled and time standards set in one month. The size of the plant will determine the number of observers, but no other system could develop a total plant-wide time standard system faster than work sampling.

Figure 11-6 shows an actual work sampling study.

QUESTIONS

1. What is work sampling?
2. What are the three techniques of work sampling, and how do they differ?
3. What is an elemental ratio, and how is it estimated?
4. Define these terms as relating to work sampling:
    a. Confidence level
    b. Accuracy
    c. Sample
    d. Randomness
    e. Sample size
    f. Probability
    g. Normal distribution
5. What are the ten steps of a work sampling program?
6. How does performance sampling improve elemental ratios?
7. What is the time standard for the following:

Job	No. Observ.	%	Hours	Units Produced
1	5,000			2,900
2	10,000			8,800
3	20,000			25,000
Idle	15,000			
Total	50,000	100%	4,250	

Add 10% allowances. What is the efficiency?

CHAPTER 12

Performance Control Systems

The motion and time study application that affects more people than any other use is performance control systems. A performance control system is very personal because it judges people, and for this reason it receives more attention than any other use of time standards. Our discussion of performance control systems is divided into three sections:

1. The functions of any control system
2. Expert opinion standards system
3. Time card system.

THE FUNCTION OF ANY CONTROL SYSTEM

Quality control, inventory control, production control, cost control, attendance control, and performance control all have the same required functions:

1. Planning or goal setting
2. Comparison of actual to goal
3. Tracking of results
4. Variance reporting
5. Corrective action.

Unless each of these functions is performed properly, there is no control system. The following discussion is to develop an understanding of what is required of each step of any control system.

Planning or Goal Setting

What is to be achieved? In every control system, the goals must be set; and most importantly, these goals must be measurable and achievable. The planning techniques used in performance control systems are the time-standard–setting techniques studied in this book:

1. PTSS
2. Stopwatch
3. Standard data
4. Work sampling
5. Expert opinion (discussed later in this chapter).

In addition to these planning and goal-setting techniques, the managers of the operation to be controlled must have an acceptable performance goal in mind:

1. With time standards and a performance control system in place, the average performance is 85%, not 100%. An expectation of much higher than 85% is not practical and chance of success is small; therefore, planning and goal setting must include standards and reasonable expectations.

2. Operators on incentive operate at 120% performance. When planning manpower requirements, 120% must be used, not 100%, or there will be too many operators with not enough work to keep them busy.

Without goals (or standards), we are without direction, and achieving our potential is impossible.

Comparison of Actual to Goal

The second function of any control system is to compare the actual results to the planned results. A quality control system would ask us to measure a product (actual results) and compare it to a blueprint (goal or standard). If the results are in tolerance, the produce is classified as a good one. Performance control systems do exactly the same.

An operator will produce a number of units in a period of time. This is compared to the number of units the time standard asked for. The result is a percent performance. Every job the operator does during a day is collected and compared to create a daily performance report. This operator is combined with all the other operators to produce supervisor, department, shift, and plant performances for each day, week, month, and year.

Tracking Results

Tracking could be called graphing. It is plotting results against a horizontal time line. In a cost control system, the goal line is the planned expenses per period. The actual expenses are plotted against this goal for each period, and any casual observer can see how the plan is going. No single period is important, but the trend or direction is important.

In performance control systems, percent performance and percent indirect are plotted weekly. The trend on percent performance (productivity) should be going up, while at the same time the indirect percentage must be held flat or reducing. The trends are what is important. Percent indirect is indirect hours divided by total hours.

Variance Reporting

When actual performance does not live up to goals (expectations), a variance from standard exists. In an attendance control system, employees are allowed to be absent twice a month without being criticized, but if they exceed this goal, a variance exists and a report would be generated by the payroll department. This is a variance report. In a production control system, if a part is behind schedule by 4 hours, it is placed on an expedite list—a variance report. Nearly every actual result will be different from the goal. The magnitude of these variances is what is important. The larger the variance, the bigger the problem. Managers should attack the largest problems (variances) first, and fix these problems.

In performance control systems, those individual jobs that reported performances lower than 70% and above 130% are collected and printed out on the variance report. Each job of the variance report becomes a project, and the reasons for the variances must be determined. Each variance can be assigned to a person for investigation and corrective action.

Corrective Action

Solving problems and implementing these solutions is the reason control systems exist. Corrective action makes it all happen. Talk to the person who is having an attendance problem and find a solution acceptable to both employee and supervisor; adjust the machine to bring quality back to standard; expedite a purchase order if material level drops below target; work overtime if behind schedule; and fix the problems causing poor performances.

Corrective action in performance control systems is wide ranging:

1. Machine maintenance problems
2. Material problems
3. Management problems such as lack of assignments, lack of instruction

4. Poor time standards

5. Poor operator effort.

Performance control systems hold problems up to public scrutiny, and problems get solved. Without performance control systems, operators know that management does not care about productivity and supervisors don't want to be bothered with problems.

Performance control systems raise the productivity of departments and plants by 42%. Taking corrective action is how that happens. A 42% decrease in direct labor cost is a significant savings, but there is no easy way. Performance control systems are hard work, but hard work is what successful people do.

EXPERT OPINION STANDARDS SYSTEM

Another technique for setting time standards is the expert opinion technique. An expert is someone who has a great experience base, and because of this experience can estimate requested work in his or her area of expertise with acceptable accuracy. I could ask you how long it takes to drive to your parents' home, and you could make a good estimate. The next time you make that trip, it will be a little longer or shorter than the estimate, but the average will probably be accurate. I could make the estimate for you using 50 miles per hour average, but I won't know the small town, the construction zones, the coffee stops, etc. required by the trip, so my standard (engineered standards) would not be as good as yours.

Many engineers, maintenance managers, office managers, and other people not on machines say, "You can't set time standards on my job." They are correct. We can't but they can! The maintenance department is a good example. As an industrial technologist, you have designed a new work station for a cost reduction. Before the cost reduction return on investment can be calculated, the cost must be estimated. You take the drawings to the maintenance supervisor and ask how much it will cost. The material cost, as well as labor cost, must be estimated, but we are only talking about labor right now. The maintenance supervisor looks at the job, mentally compares it with other jobs completed in the past, and zeros in on the estimate. The maintenance supervisor breaks down big jobs into small jobs and estimates their cost. The estimates are written on work orders and given back to you for approval. The approval process will require upper-level management's approval as well, but when the work orders are approved, they are given back to the maintenance supervisor for scheduling and completion. The maintenance supervisor enters the job on the backlog, orders material, and schedules the workers.

BACKLOG

Each service department maintains a list of jobs that have not been completed. This is the backlog, and the estimated hour total of all jobs is the backlog hours (see Figure

FIGURE 12-1 Maintenance backlog list
steps ① and ② of the step by step procedure in section 2
of this chapter

Job No.	Date Received	Job Description	Hours Required	Date Completed	Actual Hours
101	11-15-XX	Overhaul fork truck	16	11-25-XX	15
102	11-18-XX	Build safety cage	12	11-24-XX	13
103	12-1-XX	Build supervisor office	40		
104	12-3-XX	Build conveyer system	132		
105	12-10-XX	Repair machine 12y #1576-05	24		
106	12-12-XX	Build work station #1576-10	10		
107	12-12-XX	Build work station #1576-15	7		
108	12-12-XX	Build work station #1576-20	15		
109	12-12-XX	Build work station	4		
110	12-13-XX	Move machine 64	2		
ETC	50 more jobs				
		Total hours	1,324		

12-1). A backlog of work is needed to give time for planning and efficient scheduling. Management and the department manager determine what an acceptable backlog is (number of days worked) and the department manager maintains that much work.

Every job in the maintenance backlog is listed and estimated. As jobs are completed, the completion date and actual hour columns are filled in. As new jobs come in, they are estimated and added to the backlog. At the end of every week, the backlog hours are totaled and recorded on the control graph. This maintenance department has ten maintenance workers. Ten times 40 hours per week means that maintenance has 400 hours per week available. Not all work is covered by backlog:

1. Emergency maintenance has accounted for 10% of the maintenance time in the past, so 360 hours is still available.

2. Routine maintenance or scheduled maintenance is scheduled just like production work, and one of the crew spends full time oiling, changing bulbs, etc., on a routine schedule.

3. Ordinary maintenance has 320 hours per week available. How many weeks of backlog does management think appropriate? Let's say three weeks. Three weeks times 320 hours per week equals 960 hours of backlog. What should the maintenance manager do?

A Backlog control graph (figure 12-2) results from totalling the backlog list (figure 12-1) each week. The Trend line is important.

Step-by-Step Procedure for Developing a Backlog Control System

Step ①. List all jobs waiting to be completed in the department. List oldest jobs first.

Step ②. Estimate the time required for each job, and total the time. This is the beginning backlog.

Step ③. Establish a backlog hour goal. A sufficient amount of backlog is needed to allow for efficient planning and scheduling. (See Figure 12-2 step ③.)

Step ④. Establish a control graph and a control chart. Plot the beginning backlog. (See Figure 12-2 step ④.)

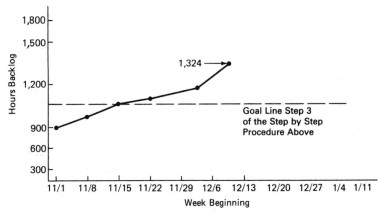

FIGURE 12-2 Backlog control graph

Step ⑤. During each week, jobs are added to the backlog, and times required are estimated. At the end of the week, the total hours added is totaled and entered onto the backlog control chart. (See Figure 12-3.)

Step ⑥. During each week, jobs completed are totaled and entered onto the backlog control chart. (See Figure 12-3.)

Step ⑦. At the end of each week, the new job hours are added to the beginning backlog hours, and the hours completed are subtracted, resulting in the ending backlog hours. This number is recorded on the backlog control graph and brought forward to the beginning backlog hours for the next week.

Step ⑧. The beginning backlog hours are subtracted from the ending backlog hours, resulting in the change or variance. Figure 12-3 is an example of a variance report.

   a. A positive variance means the backlog is growing and some action may be required.

FIGURE 12-3 Backlog control chart—variance report

Week Beginning	Beginning Backlog Hours	Hours Added	Hours Completed	Ending Backlog Hours	Change Variance
11/1	890	400	310	980	90
11/8	980	360	340	1,000	20
11/15	1,000	460	335	1,125	125
11/22	1,125	375	350	1,150	25
11/29	1,150	450	400	1,200	50
12/6	1,200	444	320	1,324	124
12/13	1,324	380	340	?	?
12/20	?				

Note: ? = fill in and plot on graph (Figure 12-2).

      b. A negative variance means the backlog is shrinking and other actions may be required.

Step ⑨. Corrective action:

      a. If the backlog trend is growing, such as in Figure 12-3, it may be necessary to add people.

      b. If the backlog trend were coming down by at least 40 hours per week, a reduction of people may be required.

      c. If the backlog were flat but too high, overtime, temporary help, or farming out work would help.

      d. If the backlog were flat but too low, loaning people out, taking vacation time, or sending people to school would correct the situation.

The backlog control system using expert opinion standards can be used for any service work such as engineering, maintenance, and office functions.

TIME CARD SYSTEM

The performance control system for most of industry is based on the individual time card. The math and the layout of forms for department, shift, and plant performance reports (daily, weekly, monthly, and yearly) are designed after the daily employee time card. The time card is a different form than the weekly payroll card and is punched upon arrival and leaving the plant. This time card is only for performance control. Figure 12-4 is an example of a completed time card.

Step-by-Step Procedure for Time Card Calculation

Steps ①–⑦ are completed by the operator.
Steps ⑧–⑬ are completed by the time keeper.

FIGURE 12-4 Time card example

shift is 7 A.M. to 3:30 P.M.
lunch 11 A.M. to 11:30 A.M.

Operator ① Name: Mary				② Date: 12/13/XX			③ Department: Press		
Time In	Time Out	Job No.	Oper. No.	Time Std.	Pieces Produced	Actual Hours	Earned Hours	%	Indirect Time
7:00	8:45	1660	10	100	200	1.75	2.00	114	
8:45	10:15	1700	15	167	250	1.50	1.50	100	
10:15	12:15	1660	15	850	1,500	1.50	1.76	117	
12:15	1:00	Meeting				—	—	—	.75
1:00	2:06	1750	15	45	55	1.10	1.22	111	
2:06	3:30	1800	10	750	1,000	1.40	1.33	95	
Total						7.25	7.81	108	.75
④	④	⑤	⑤	⑥	⑦	⑧ ⑫	⑨ ⑫	⑩ ⑫	⑪ ⑬

Step ①. Name: The operator's name goes here. In large systems or computer systems, an operator number may be used, but for good employee relations, the name should always be included.

Step ②. Date: The date the work is performed goes here. As always, the complete date should be used.

Step ③. Department: Each employee is assigned to a department, and all employees of a department are combined together for a department performance report. If an operator works in two departments on a given day, two time cards are used. Department numbers are often used instead of department names.

Step ④. Time in, time out: A time clock could be used, but having the operators write the time in is the preferred practice. Decimal hours are preferred, and six minutes is .1 hours. Well-conceived systems allow only one column for time in, and the time in on one job is the time out on the previous job. This eliminates gaps in time that are impossible to figure out after the operator goes home.

Step ⑤. Job no. and operation no.: Having the operator fill in the job number and operation number ensures that the operator knows the correct time standard. The information can be retrieved from a traveling route sheet or operating instructions at the supervisor's desk. The time keeper also looks up these numbers to check the time standard. The operator must know what is expected, or we have no performance control system.

Step ⑥. Time standard: This is the engineering time standard in pieces per hour. The hours per unit or 1,000 units is sometimes used, but pieces per hour is easier for the operator to use.

Step ⑦. Pieces produced: The number of units made during the run of the job goes here. The count can come from a counter mounted on the machine or work station, or it can be predetermined by standard container counts. Pieces produced always means good pieces. Scrap cannot be allowed in production figures or efficiency.

The operator is required to fill in all the information up to this point. Time keepers should audit this information just like any good management system.

Step ⑧. Actual hours: The actual hours are the number of clock hours the operator actually works on the job. Mathematically, actual hours are the ending time less the beginning time. A normal day is 8 hours, and the total of actual hours should be eight. (More on this in Step ⑫.)

Step ⑨. Earned hours: Earned hours are the amount of work the employee did compared to the time standard. Earned hours are calculated by dividing the time standard (Step ⑥) into the pieces produced (Step ⑦).

Step ⑩. %: Percent performance is the relationship between actual hours and earned hours. If the operator earned more hours than used, the percent performance is over 100%. If the operator earns fewer hours than actually used, the percent performance is under 100%. Percent performance is calculated by dividing earned hours (Step ⑨) by actual hours (Step ⑧).

Step ⑪. Indirect hours: Indirect hours are hours worked that are not covered by the time standard. Indirect hours are kept separate from direct labor hours. Indirect hours are neither in the actual hours nor the earned hours, but are recorded in this column. Indirect labor hours must be controlled by reason code. Each indirect labor charges have a reason code number, and time is totaled and summarized by code number. Example reason code numbers:

01 Maintenance
02 Quality
03 Safety
04 Material handling
05 Clean-up
06 Set-up
07 Rework
08 Miscellaneous

The number of indirect codes can be large. Each cause of indirect labor must be controlled, so inflation does not occur. Achieving 100% performance would be easy if production were run while charged out on an indirect code. In well-managed plants, indirect labor is kept under 20% of total labor.

Step ⑫. Totals: Total actual hours, the total of column ⑧, is divided into the total earned hours (Step ⑨) to calculate total percent performance (Step ⑩). The only way to calculate percent performance (whether for a job, a person, a department, a plant, a day, or a year) is to divide total actual hours into total earned hours.

Step ⑬. Total indirect hours: Total indirect hours are added to total actual hours to figure how many total hours were paid for. (Total indirect hours are discussed further later in this chapter.)

Step ⑭. The time card hours are compared to the payroll hours. These two numbers must come out the same. This process is called justifying time card and payroll hours.

Figure 12-5 show examples of daily performance reports. The step by step procedure that follows figure 12-5 shows how to calculate department performances.

Step-by-Step Procedure for a Department Performance Report (Daily)

Step ①. Daily performance report—Dept.: One performance report per department and one summary for all departments is needed. If the plant has ten departments and 200 people, there will be eleven daily performance reports—one for each

FIGURE 12-5 Performance control system: Daily department reports, one for each department plus one for the plant total.

Daily Performance Report   *Assembly*

Daily Performance Report   *Dept. Paint*

Daily Performance Report   *Dept. Press*

Date 12/13/XX   *Supervisor's Name Dale* ③

Operator Name	Actual Hours	Earned Hours	%	Indirect Codes								Σ	%
				01	02	03	04	05	06	07	08		
Mary	7.75	7.81	108								.75	.75	9
Fred	8.0	9.25	116								—	—	
Jim	1.25	1.50	120		4				2		.75	6.75	84
Bob	—	—					6	2				8	100
Lynn	8.0	10.0	125										
Tommi	7.25	8.00	110								.75	.75	9
Pat	8.0	12.0	150										
Bill	—	—		3				2	2.25	.75		8	100
Dee	8	6	75										
⑪ Total	47.75	54.56	114	3	4	—	6	4	4.25	.75	2.25	24.25	34
④	⑤	⑥	⑦				⑧					⑨	⑩

department supervisor and one for the plant production manager. The plant performance report would look just like the one in Figure 12-5, except with supervisors' names where the employee names are.

Step ②. Date: The date the work was accomplished.

Step ③. Supervisor's name: The department performance report measures the supervisor's performance. Productivity performance is one of the top two measures of production management, although there are many more.

Step ④. Operator names: The time keeper maintains preprinted daily performance reports with operator names already included. The time keeper will put the time cards in name order, and then transfer the information for columns ⑧, ⑨, ⑩, and ⑪ onto the daily departmental form.

Step ⑤. Actual hours ⎤ These three columns are transferred

Step ⑥. Earned hours ⎬ from the time card totals to the

Step ⑦. %   ⎦ daily performance report as they are.

Step ⑧. Indirect codes: Indirect time charged on individual time cards is coded and listed to the right of the operator's name.

Step ⑨. Σ: The sum of each individual's indirect hours is placed here.

Step ⑩. %: The percent indirect hours is calculated by dividing the indirect hours by the total hours worked (direct actual hours plus total indirect hours).

Step ⑪. Total: Each column is totaled, except the two % columns.

    a. % Direct Step ⑦ is calculated by dividing earned hours by actual hours.

    b. % Indirect Step ⑩ is calculated by dividing hours by actual hours plus indirect hours.

The daily performance reports for operators and supervisors are good indications of how their days go, and a good work record is built on many good days.

Weekly Performance Reports

Weekly performance reports (see Figure 12-6) are exactly the same as daily reports, except the day of the week replaces the operator's name and daily totals are placed in each block. Notice in Figure 12-6 that the information for Mon. 12/13/XX has been transposed from the bottom of the daily performance report Figure 12-5. Not one change has been made. At the end of the week, each column will be totaled except the two percentage columns (just like the daily report), and the percent direct labor and the percent indirect are calculated just like the daily performance report. Notice how the columns in Figures 12-5 and 12-6 line up exactly.

There will be the same number of weekly, monthly, and yearly reports as there are daily reports. The time it takes to calculate any of these reports by hand is only a couple of minutes per time period because only transferring information is required.

FIGURE 12-6 Performance control system: Weekly performance reports, one for each department plus one for the plant total.

Day	Actual Hours	Earned Hours	%	Indirect Codes								Σ	%
				01	02	03	04	05	06	07	08		
Mon.	47.75	54.56	114	3	4	—	6	4	4.25	.75	2.25	24.25	34
Tue.													
Wed.													
Thurs.													
Fri.													
Sat./Sun.													
Total													

Weekly Performance Report Dept. *Press*
Week ending: *12/18/XX* Supervisor's name: *Dale*

The monthly report will have week ending dates instead of days of the week. The yearly report will have months instead of days of the week. Everything else is the same. On a yearly report, twelve months of information is available on one $8\frac{1}{2} \times 11$ sheet of paper per department. This information becomes valuable in planning budgets for next year.

Tracking

A graph of weekly performances will be useful to management. If anything is going right or wrong with productivity, the tracking will highlight it. Three pieces of information should be tracked (see Figure 12-9):

1. Percent performance
2. Percent indirect
3. Output per manhour.

Percentages can be measured on the left side of a graph, while output per manhour can be measured on the right side. The output per manhour is a good check of the percent performance accuracy. Both should go up and down at the same time if time standards are meaningful.

Variance Reporting

The time cards can be returned to the operator once the daily performance report is calculated. If an operator's performance is out of normal, the supervisor can talk to the operator to find out what happened. The goal is to correct problems. The operator is the best source of information, but the operator should notify the supervisor when having

FIGURE 12-7 Time card problem
shift 7 A.M. to 3:30 P.M.
lunch 11 A.M. to 11:30 A.M.

Time In/Out	Job No.	Oper. No.	Time Std.	Pieces Produced	Actual Hours	Earned Hours	% Performance
7:00	1600	20	125	300			
9:00	1610	25	1,200	2,000			
10:45	1500	15	45	120			
2:00	1700	25	300	600			

trouble. Often the operator does tell the supervisor of a problem, but the supervisor forgets to assign someone to fix the problem. Supervisors often are bombarded with problems from many directions at the same time, and forgetting one problem is easy.

Computer programs are also used to highlight variances that are beyond acceptable limits. Limits must be set, and variance beyond the limits will be listed. For example, any work less than 70% or over 130% performance needs to be reviewed. If the com-

FIGURE 12-8 Example: Actual performance report

Crain Enterprises, Inc.
Plant Performance Report

Employee	02-Mar	09-Mar	16-Mar	23-Mar	30-Mar	07-Apr	14-Apr	21-Apr	28-Apr
Bailey	N/A	N/A	N/A	N/A	N/A	N/A	N/A	N/A	N/A
Baird	77	90	99	78	99	84	104	97	110
Cook	52	63	95	83	94	47	56	53	90
Cross	73	95	95	107	VAC	85	91	92	95
Curley	N/A	N/A	N/A	N/A	N/A	N/A	46	63	66
Dover	N/A	128	116	196	87	160	112	N/A	140
Goins	107	105	118	128	115	120	128	111	99
Goins	85	107	107	74	83	100	112	108	102
Hannan	140	137	120	115	108	112	140	115	118
Isom	111	106	98	94	87	107	118	90	100
Jones	66	68	65	65	84	103	93	101	91
Kennedy	66	75	93	95	101	120	116	120	127
Marrs	74	100	111	82	84	90	103	82	91
Monan	87	95	100	N/A	N/A	N/A	N/A	86	87
Polley	VAC	128	124	116	115	113	119	119	123
Stigall	N/A	141	111	131	115	121	129	101	101
Taylor	112	135	130	123	129	126	150	131	109
Taylor	85	104	95	69	74	91	127	104	99
Wilson	107	131	130	130	97	131	132	130	128
Plant Total:	87	104	105	97	98	108	110	100	103
Percent Direct Hours	75	75	69	57	59	62	66	67	66

Week Ending/Percent Efficiency

puter lists variances in order of magnitude, the worst problems will be on top of the list.

Corrective Action

The reason for variance reporting is to identify the problems so they can be fixed and eliminated. Each variance becomes a project, and a project engineer can be assigned. The project engineer is the person who can best fix the problem:

1. Industrial technologist
2. Supervisor
3. Maintenance person
4. Operator
5. Tooling manager, etc.

A project control system is needed, and the most important aspect of this system is to motivate the project engineer to get the job done. The project list should be distributed to all engineers, managers, and the top manager every week.

					Percent Efficiency							
05-May	12-May	19-May	26-May	02-Jun	09-Jun	16-Jun	23-Jun	30-Jun	07-Jul	14-Jul	21-Jul
N/A	N/A	N/A	N/A	N/A	N/A	69	85	95	CLOSED	97	113
103	100	104	127	150	73	N/A	N/A	N/A	CLOSED	N/A	N/A
90	89	81	91	119	118	114	N/A	N/A	CLOSED	N/A	N/A
87	82	121	106	112	103	104	102	104	CLOSED	107	122
104	118	115	118	123	LOA	LOA	LOA	108	CLOSED	123	109
N/A	115	127	123	VAC	131	112	117	96	CLOSED	114	132
110	116	128	124	119	96	122	VAC	121	CLOSED	119	113
99	132	125	138	134	96	104	110	117	CLOSED	110	116
128	140	142	110	127	130	130	135	147	CLOSED	123	136
99	103	104	108	107	117	110	111	98	CLOSED	N/A	N/A
101	107	124	117	109	108	108	80	107	CLOSED	114	108
VAC	97	119	109	144	123	137	104	130	CLOSED	136	137
82	112	114	69	83	81	86	98	88	CLOSED	IND	93
104	123	121	106	107	97	85	VAC	91	CLOSED	94	119
114	113	120	114	126	110	130	118	118	CLOSED	102	107
N/A	118	120	128	124	121	113	LOA	106	CLOSED	122	104
131	130	118	122	114	121	135	93	112	CLOSED	125	123
100	122	127	119	107	VAC	118	117	122	CLOSED	118	111
121	N/A	139	VAC	138	138	133	112	133	CLOSED	142	145
104	112	121	115	121	112	113	106	114		116	120
57	58	70	62	60	65	73	74	65		76	69

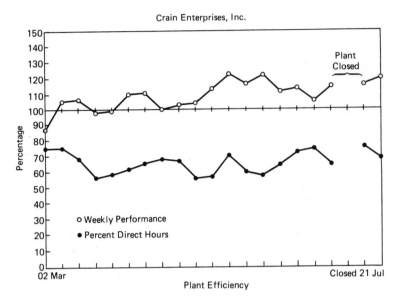

FIGURE 12-9 Example: Actual performance graph (same data as in
Figure 12-8)

Conclusion

A properly designed and operated performance control system can save thousands of
dollars per week. The productivity of the plant increases an average of 42%, and cost
decreases proportionately. Manufacturing and distribution companies cannot afford not
to have performance control systems.

Figures 12-8 and 12-9 show actual performance reports and control charts.

QUESTIONS

1. Why is a performance control system given so much attention?
2. What are the five functions of any control system?
3. What is proper expectation for productivity?
4. What percent performance will you achieve if you tell the operator that 85% performance is acceptable?
5. What is an expert opinion standard?
6. Who could use expert opinion standards?
7. What is a backlog? Why do we need a backlog? How big should a backlog be?

8. What are the nine steps of a backlog control system?

9. What is the percent performance for the time card problem shown in Figure 12-7?

10. What is the relationship between individual time cards and daily, weekly, and annual performance reports?

CHAPTER 13

Wage Payment Systems

Wage payment is a measure of value; therefore, it is more important than mere economics would indicate. People are classified as hourly or salaried and exempt or nonexempt. People are paid by the hour, week, month, or by the piece, and they can earn extra money through bonus plans, commissions, and profit sharing.

This chapter builds on the performance control system of the previous chapter and time standards developed throughout this book. So far, we have developed the goals and reported the percent performance, but how much do we pay people? There are three ways of paying employees:

1. Salaries: hourly, weekly, or monthly
2. Incentive and commission
3. Bonus and profit sharing

SALARIES: HOURLY, WEEKLY, OR MONTHLY

How much is each position worth per hour, week, or month? To answer this question, a job evaluation must be made for all positions. Each position earns point value: The higher the point value, the more complicated the position. Groups of positions with similar point values are pulled together into a job classification. Once we know the relative value on these job classifications, a local wage and salary survey is used. Management must make a decision as to the wage percentile in which its company will compete. It may choose to pay the average rate, or above average, or even below average, but once the percentile has been determined, the rates are set using that percentile from the wage and salary survey. A salary range may be specified to allow for longevity and experience.

Exempt versus Nonexempt

An exempt category of employee means that these employees are exempt from the wage and hour laws. Supervisors, engineers, and managers are in this classification and have no protection under the law. A nonexempt employee will be paid overtime for work over 8 hours per day and 40 hours per week, must be paid at least the minimum wage, and cannot be discriminated against because of race, religion, sex, national origin, or age. Exempt employees are salaried and are often paid a fixed sum per week or month, no matter how many hours are involved. Nonexempt employees can be either hourly or salaried but are paid for overtime on an hourly basis.

A rate of pay has been calculated for each position, and a time card is legally required for every nonexempt person. The time clock hours multiplied by the rate per hour is the weekly pay. This is true of hourly or salaried nonexempt employees. Exempt employees are automatically paid until terminated.

Measured Day Work

A measured day work system is a performance control system based on time cards and time standards. The time card has nothing to do with payroll unless it is used in combination with an incentive system. Measured day work is for productivity measurement and control and will generate a large savings in labor. About 75 to 85% of all factory employees use job time cards for measured day work. Measured day work is one step required before starting an incentive system and was illustrated in chapter 12 as the performance control system.

Day Work

The term *day work* refers to being paid by the hour, week, or month. Day work has no time standards or performance calculation. Day work pay systems are, in fact, no system at all.

U.S. Government Bodies that Regulate and Influence Wages and Salary Practices

1. Wages and hours division, Department of Labor: administrates minimum wage, overtime, etc.
2. National Labor Relations Board: regulates labor practices
3. Equal Employment Opportunity Commission: investigates charges of discrimination.

Job Evaluation Technique

Job evaluation for wage and salary determination is the process of determining the value of each specific job or position in a company and comparing this value to every other job or position in the company. The purpose is to set wage and salary levels for all employees.

During World War II, the following organizations worked together to design a fair and equitable method of determining wages and salary:

1. National Electrical Manufacturers Association

2. National Metal Trades Association

3. General Electric Company

4. Westinghouse Electric Company

5. U.S. Steel.

The results of their work were like the factors and point value shown in Figure 13-1. A description of each degree would be helpful to make proper assignments of point values for each factor, but this is a subject for an advanced course.

A technologist making the evaluations must have the following information:

1. A good understanding of all jobs

2. A good knowledge of the company

3. Direction from top management

FIGURE 13-1 Wage and salary point system

Factors	Degrees and Point Values					
	1	2	3	4	5	6
1. Work experience needed	22	44	66	88	110	120
2. Essential knowledge and training needed	14	28	42	56	70	90
3. Initiative and ingenuity	14	28	42	56	70	
4. Analytical ability	30	40	50	60	70	
5. Personality requirements	20	30	40	50	60	
6. Supervisory responsibility	5	10	15	20	25	
7. Responsibility for loss	10	20	30	40	50	
8. Physical application	10	20	30	40	50	
9. Mental and visual application	5	10	15	20	25	
10. Working condition	10	20	30	40	50	
11. Dexterity	60	120	180	240		
12. Character of supervision required	10	20	30	40	50	
13. Responsibility for confidential matters	5	10	15	20	25	

a. Number of wage classifications

b. Company's wage percentile goal

4. An area wage and salary survey.

Once armed with this information, the six-step guide to making a job evaluation can be used.

Steps in Making a Job Evaluation

1. List all the jobs to be evaluated on a single sheet of paper. Make fourteen copies of this list—one for each *factor* and one total sheet.

2. Assign the number value for each job, one factor at a time. This will allow you to think about only the work experience required for each job, thereby doing some ranking and grouping of jobs. Go through all thirteen factors, but do every job for one factor at a time.

3. After all factors for all jobs have been evaluated, use the fourteenth copy of the job list to total all the numerical values for each job. The higher the point value, the more complicated the job.

4. List all the jobs starting with the highest point value first, down to the lowest point value last.

5. Group the jobs into pay grades—depending on how many ranges your management has recommended. For example, some firms have only four different wage rates; others have as many rates as they have people. It is better to have as few as possible to eliminate excessive job bidding and transfer.

6. Compare your wage rate with the community's and recommend a wage and salary plan.

Some words of caution for making job evaluations:

1. Do not think about the people in the jobs now. Think only of job content.

2. Make your pay grade groupings logical. After totalling all the points for each job, we list those jobs and points in reducing order. Look for large gaps in the totals and pick the midpoints of these number gaps as your pay grade breaking points. One point should not be the difference between two pay grades because we are not that accurate. For example, if the total points for several jobs were 300, 299, 297, 280, 279, 277, 275, we would pick 288 as the breakpoint between pay grade 1 and pay grade 2.

3. Do not make the evaluations public knowledge, or you will spend the rest of your career justifying your position.

4. If management recommends ten wage scales and nine or eleven would be better, tell them.

INCENTIVE SYSTEM AND COMMISSIONS

Incentive systems are as old as humanity, and they work because they satisfy the human needs of survival, recognition, and ego gratification. (See Figure 13-2.) They are not used often enough because they are a lot of work.

A properly designed incentive system will

1. Reduce unit cost
2. Increase equipment utilization
3. Promote the competitive spirit
4. Increase employee pay for increase effort
5. Improve job satisfaction
6. Recognize outstanding employees.

Every manager has seen what can happen when one employee really cares about his or her job. That employee is usually promoted or given other recognition and rewards. Just think what could happen if a whole plant cared. An incentive wage plan will do this.

The National Science Foundation found that when workers' pay is linked to their

FIGURE 13-2 Good incentive systems are equally good for employees and the company.

performance, the motivation to work is raised, productivity is higher, and the employees are likely to be more satisfied with their work.

A Mitchell Fein Study of 400 plants in the United States found that incentive systems increased performance by 42.9% over measured day work systems, and 63.8% over plants with no standards.

Managers in the United States believe in incentives for managers. Eighty-nine percent of manufacturing companies have executive bonus plans, and they earn an average of 48% over their base pay. Even more important is that their companies earn an average of 43.6% more pretax profit. Yet only 26% of U.S. labor has incentives.

Sales management uses incentives more than any other area of business. Incentives in sales are known as commissions, and salespeople usually receive a minimum salary as a base plus a percentage of their sales. Sales commission incentives have been very successful, and no one would consider eliminating them, so why do managers reject the fact that incentives will work in operations?

A Bureau of Labor statistics study in 1976 of 1,711 union contracts (7.6 million workers) showed that 1,504 contracts allowed time standards (5.5 million workers) and 476 contracts allowed incentives.

Types of Incentive Systems

Incentives can take many forms:

1. Percent of sales (commission)
2. Piecework and differential piece rate (Taylor piecework)
3. Earned hour plans, standard hour plan
4. Time off with pay
5. Productivity sharing, Improshare, and Scanlon
6. Suggestion system
7. Profit sharing, bonus systems.

Each plan is discussed in detail later in this chapter, including the advantages and disadvantages. There is an incentive plan for every situation.

Goals

Incentive systems design should start with a good list of goals and philosophies so everyone will understand why the system was implemented. Goals of an incentive system are to

1. Reduce cost (must be first and foremost)
2. Increase productivity—produce more with less
3. Increase employee earnings

4. Improve employee morale

5. Improve labor/management relations

6. Reduce delays and waiting time

7. Improve customer service

8. Develop motion consciousness and cost consciousness

9. Reduce need and type of supervision

10. Increase plant and machine utilization.

Philosophies

Philosophies spell out the way we want the system to work and are used to direct future discussions and problem resolution. For example,

1. No one will be laid off because of the incentive system.

2. Extra pay for extra effort—120% is normal.

3. No credit is earned for producing rejects or scrap.

4. Increased wages can only be earned for work done on standard. No average pay for nonstandard work.

5. No time standard will be changed unless the job changes, and time standards will be kept current.

6. Complaints will be answered as soon as possible.

7. High earnings by employees are good because of savings in overhead and fringe benefits.

Potential Problems

There are some problems to be aware of when using any incentive system. These problems should be understood to keep from making errors in the design of a new system:

1. Motion and time study technicians are needed to set and maintain time standards and to investigate problems. This cost can be from 1 to 3% of the production crew being covered.

2. Time standards that are not kept current will lead to increased cost.

3. Union grievances will increase. It will happen, so be prepared.

4. Management must put more effort, manpower, and cost into

   a. Quality control

   b. Production control.

5. If time standards are too tight, the employees will lose motivation. Average earnings are 120%, and each person in the plant must have the opportunity to earn a 20% bonus.

These problems must be recognized and communicated to top management so realistic expectations are set in their minds. These problems will in no way cost anywhere near the savings potential of an incentive system, but if top management is surprised by any of the increased problems, the technologist will have trouble.

Individual Incentive Systems

Individual incentives are greater motivators than group incentives. Employees feel they have less and less control with the growth of the group size. A good rule of thumb is to use individual incentives wherever you can. Group incentives should be used on assembly lines, multiple operator equipment, and other team work where individual initiatives cannot be used. Keep the groups as small as possible. The following list and discussion of individual incentive systems is limited to only the most popular and usable. The subject continues to grow, and the technologist must keep up with developments.

Commissions Commissions are used extensively in sales. The salesperson will get a percentage of the sales dollar as his or her salary. Salespeople will have a minimum salary called a base, but this salary (if indeed there is one) would not sustain the person to any desired level. If the salesperson wants a better lifestyle, he or she must go out and work for it. Most salespeople would not want it any other way, and most of us recognize that salespeople of our organization are the top earners. Most of management recognizes the pressure put on salespeople to produce (some of them envy the salespeople), but no one would suggest removing the commissions.

A type of commission called royalties is used for other industries such as publishing, film making, and inventing. With royalties, individual effort can produce increased income. The employee gets a commission depending on the success of the mission. Royalties are effective because motivation to succeed is great. Armies once used to share the looted goods they took from the defeated enemy.

Other examples of commissions or royalties include the following: accountants and lawyers earn a percentage of the billable hours; a barber earns a percentage of his or her gross income; a sharecropper keeps a percentage of the crops; and any serviceperson could be awarded a share of the income to motivate him or her to produce.

Piecework An employee earns a specific dollar per unit produced. In farming, a picker earns so many dollars per pound, box, or bag. A fishing crew earns so many dollars per pound of fish. If these people don't produce anything, they earn nothing. In manufacturing, an employee is guaranteed an hourly rate until the standard pieces per hour is reached.

Straight Piecework After the standard pieces per hour is reached, the earnings are increased at the unit rate.

$$\text{Unit rate} = \frac{\text{(wage) \$7.00/hour}}{\text{(standard) 100 pieces/hour}} = \$.07 \text{ per unit}$$

For example, if in 8 hours the person produced 1,050 units, the employee would earn $73.50 (.07 × 1,050) for the day. The worker would never earn less than $56.00 (8 × $7.00), the base rate.

Each part has a unit rate. This has been a problem in the past, before computers, when the base rate changes (every unit rate would change). Therefore, management developed the concept of cost of living adjustment (COLA). The incentive pay would continue on the original base rates, and the COLA was added to the pay at the end of the pay period. However, after many years the COLA grows and the base rate becomes a small part of the hourly rate. The incentive to work becomes less and less as the COLA grows. An example is a base rate of $2.50/hour plus $5.00 per hour COLA. The incentive is based only on the $2.50 per hour rate instead of $7.50. The incentive is only one third of what it should be.

Differential Piece Rate This piece rate system uses two rates:

1. Up to 100%; a smaller unit rate
2. Over 100%; a larger unit rate.

This incentive plan tends to be more effective as a motivator than straight piecework, but it never gained the popularity of the straight piecework plan because of its complexity.

Taylor Multiple Piecework Plan Taylor's plan was to discourage poor performers while attracting and keeping high performers. No minimum salary was provided, and only 50% of the unit rate was paid until 100% was achieved. Over 100% was paid at the rate of 125% of the unit rate. This large disparity created a strong incentive to produce. Weak and marginal employees soon left the company. This plan would not be acceptable today, but it is still interesting.

Earned Hour Plan or Standard Hour Plan These are the same plans, and they are the most popular incentive plans used in manufacturing today. The earned hour plan is built on the performance control system discussed in chapter 12. The earned hours plus indirect hours multiplied by the employee's current hourly pay rate equals the day's pay.

These plans guarantee that the employee will never earn less than the base rate, and for every percent over 100%, a 1% bonus is earned. No incentive is earned on time spent off standard (clean-up, material handling, waiting), but no penalty is charged either. Employees soon learn that the only way to earn an incentive is to stay on jobs that have time standards.

Example: An employee works 7 hours on a job with a time standard of 250 per hour. The operator produces 2,000 units and spends one hour on preventive maintenance. At the rate of $7.00/hour, how much does this person earn?

$$\frac{2{,}000 \text{ produced}}{250 \text{ standard}} = 8 \text{ earned hours}$$

Plus 1 hour maintenance
9 total hours pay due
Employee's present salary = $ 7.00/hour
$63.00/day

$$\% \text{ performance } = \frac{8 \text{ hours earned}}{7 \text{ hours actual}} = 114\%$$

Look back at the time card problem in chapter 12 (Figure 12-7). Does this employee deserve a bonus? The performance control system described in chapter 12 is the technique most commonly used to pay incentive wages.

Time Off with Pay Additional production may not be needed, and instead of laying off part of the work force and paying the remaining employees a bonus, the company lets the employees go home once the goal has been achieved. This technique was used at an oil company. This company had produced two million gallons of oil packaged in quart cans using the wrong formula. The oil was recalled and sent to a newly rented warehouse, where it was to be removed from the cartons and cans. The oil was sent back to the refinery in tank cars, the cardboard was scrapped, and the cans were sold as scrap metal. A production line was designed, built, and manned. The production standard was set at 2,000 cases per shift. The average after a month was 1,200 cases per shift, and 1,500 cases was the best day. The technician was absolutely convinced that 2,000 was a good standard and sold management on the first incentive system ever used at this company. The technician worked with the crew on the first day of the program, and 2,000 units were produced in $7\frac{1}{2}$ hours. The crew went home early, and the supervisor and the technician cleaned up the mess. The crew knew they could achieve the standard. From that day on, the crew finished early and cleaned up the plant. Before the end of the project, the employees were going home two hours early and receiving pay for 8 hours. The company got a 42% increase in production, and the employees got 2 hours of additional pay without putting in the time. Everyone can be a winner in a good system.

Group Incentive Systems

Productivity sharing plans are a partnership entered into by labor and management to share in any savings produced by working smarter and harder. Productivity sharing plans are group plans and promote team work. Everyone can participate in the incentive bonus—hourly, salaried, exempt, nonexempt, clerical, craftsmen, etc.—and they can share equally. There are many plans, but I discuss only two.

Improshare by Mitchell Fein and Associates Improshare measures productivity as it has been over the past year using common output per manhour figures. This is called the base period, and any improvement over the base period is shared 50/50. Fifty percent of the savings are split by the employees, and 50% of the savings stay in the company to reduce cost, reduce price, or increase profits.

An example of this technique is a fiberglass bathtub/shower stall plant that produced 100 units per day with fifty total people. The only person not in the fifty-people count was the plant manager. Everyone in this plant knew that over 100 units was a good day and under 100 units was a bad day. The measure of productivity was hours per unit:

$$\frac{50 \text{ employees } \times 8 \text{ hours/day}}{100 \text{ units}} = 4 \text{ hours per unit.}$$

The evening before the start of the incentive program, the employees were told that anything they could do to reduce the hours per unit figure would be shared with them 50/50. They were told that the standard would not be changed unless an expense of over $10,000 was required to make productivity improvements, and then the standard would only change by 80% of the normal change. The 20% left in the standard would allow the employees to share the improvement, but the company needed the reduced cost to remain competitive.

The advantages of Improshare are as follows:

1. The standards are set to the present production output—not engineering standards.
2. Savings can start immediately.
3. Everyone participates. Everyone is on the same team.
4. Everyone gets the same bonus, and it is based only on the the number of hours worked.
5. Employees become equal partners with the company on productivity gains.
6. Improshare is based only on quality output.
7. The program is easy to set up and easy to maintain.
8. A production employee can make the calculations.

The calculations are easy, as shown in Figure 13-3.

Complete Friday, Saturday, and total. How much does every person take home per hour?

The math can be explained as follows:

1. Units produced is counted as finished parts are moved into the warehouse. Only good units are moved into the warehouse, and if a customer sends one back, it is subtracted until it is reworked and sent back to the warehouse as a first-quality unit. A big cost improvement was realized by the company when seconds (not quite first quality) were eliminated.

2. Standard hours per unit are always the same.

FIGURE 13-3 Weekly Rockville plant incentive plan

	Mon.	Tue.	Wed.	Thur.	Fri.	Sat./Sun.	Total
1. Units produced	100	110	100	110	115	—	—
2. × stand. hours/unit	4	4	4	4	4	4	4
3. = earned hours	400	440	400	440	—	None	—
4. − clock hours	408	400	384	384	376	16	—
5. = bonues hours	—	40	16	56	—	—	—
6. × 50%	—	20	8	28	—	—	—
7. × avg. labor rate	7.00	7.00	7.00	7.00	7.00	7.00	7.00
8. = employee bonus $	—	140	56	196	—	—	—
9. Divided by clock hours = $ bonus/hour	—	.35	.15	.51	—	—	—

3. Earned hours equal units produced times the standard hours per unit (line 1 × line 2).

4. Clock hours are the actual time card hours. If fifty employees were there all 8 hours, then 400 hours would be paid. A time keeper would provide this figure, and it is one of the most reliably kept statistics in all of industry.

5. Bonus hours: If the clock hours are less than the earned hours, a bonus is earned. Negative numbers do not need to be extended because no bonus is earned and no pay will be subtracted (line 5 = line 3 − line 4).

6. Fifty percent of the bonus hours are the employees' bonus. Fifty percent of line 5 = bonus hours.

7. Average labor rate is calculated monthly by adding up all the base hourly wage rates and dividing by the number of employees. It is simply the arithmetic average hourly wage rate. It is constant for at least one month, and it doesn't change much.

8. Employee bonus equals the bonus hours (line 5) times the average wage rate (line 7). This is the money to be divided among the employees. To make everyone equal partners in the incentive plan, hours worked is the only reason for a bonus to vary.

9. Bonus/hour equals the employee's bonus (line 8) divided by clock hours (line 3). The decimal in two places means cents (.35 = 35 cents).

The potential of Improshare is outstanding. Team development and idea sharing are key to high earnings. Management of a plant using the Improshare system changes from a pusher/controller to a problem solver/implementer of ideas. Improshare leads to some interesting question:

1. Who pays for overtime for maintenance?
2. Who decides if we replace the person who quit?

3. What happens if management doesn't implement our suggestion?

4. What do we do with the person who is not pulling his or her weight?

Improshare is a system where everyone can win.

Scanlon Plan An older group productivity sharing plan which involves the work force in the cost reduction efforts is the Scanlon plan, developed by Joe Scanlon in 1929. Three factors must be in place for a Scanlon plan to work:

1. Bonus payment

2. Identity with the company's problems

3. Employee involvement

Like Improshare, the Scanlon plan starts with a measure of productivity. In Scanlon, it is the base ratio:

$$\text{Base ratio} = \frac{\text{payroll cost of all involved}}{\text{value of production produced}}.$$

As an example, let's say a 110-person company has a weekly payroll of $44,000 and has sales of $550,000 per week.

$$\text{Base ratio} = \frac{44,000}{550,000} = .08$$

The ratio has been 8% during the previous several months. If this week our payroll was $42,500 and we produced $600,000 worth of product, our labor budget would have been $48,000 (8% of $600,000). We only used $42,500, a savings of $5,500. This savings would be shared with the employees according to some plan like Improshare's 50/50; therefore, each of the employees would receive a portion of $2,750.00.* Any method of sharing must be fair to all, and number of hours worked seems to be the best. If payroll said we used 6,071 hours last week, then the employees' share of the bonus pool would be divided by this number.

$$\frac{\text{Bonus dollars}}{\text{Hours worked}} \quad \frac{\$2,750.00}{6,071} = \$.45/\text{hour}$$

Each employee will receive $.45 for each hour worked last week.

Quality circles or employee/management meetings must be frequent. The meeting will generate ideas to improve production efficiency. The savings from these ideas are shared with all employees.

Continued company training programs are needed to communicate the company's needs and problems.

The Scanlon plan could be expanded into the cost-of-goods-sold concept; and any

---

*$2,750.00 is 50% of the total savings of $5,500.00.

material, supplies, or other overhead expenses could be fare game for cost reduction. This would make managers out of every employee.

Suggestion Systems

Suggestions were a part of the previous two incentive systems, but they could be used with any pay plan. Suggestion systems can be the entire incentive plan or one part of an individual incentive plan. A typical suggestion system will say, ''The company will pay the employee whose suggestion has been accepted one half of the first year's savings.'' This can be considerable money. There are many variations, but 50% of the first year's savings is normal. Large savings may be paid in three phases: first payment due upon acceptance by management; second payment made upon successful implementation; and final payment after six months, if the suggestion truly saved the amount claimed. This payment system can help the employee spread the tax liability.

Suggestion systems give incentives for creative ideas. If an employee has an idea that will save time and effort, what motivation does he or she have to give this idea away? If on incentive, the employee can earn more by keeping the idea to himself or herself. If not on incentive, the idea helps only the company. Sure, it will reduce cost and save jobs, but this doesn't mean much to the employee.

Some successful suggestion systems have been based on standard awards of $10, $25, and $100. This system is based more on the reward of recognition than the reward of money.

Suggestion systems are many, but not too many are considered successful. Success is measured by the number of suggestions submitted and implemented. One reason for the low number of suggestions is the reluctance of production employees, including supervisors, to put anything in writing. A company should encourage employees to bring their ideas to a person who has some ability to write.

If an idea is rejected, management must take great care in explaining to the submitter why the idea will not work. Every effort should be made to implement employee suggestions. Any ties or close calls should be implemented; otherwise employees will get discouraged.

Winners of cost reduction awards should be asked if they want to make their good fortune public. Many employees worry about the social pressure resulting from helping management. They may wish to tell only a few close friends.

Bonuses and Profit Sharing

Bonuses and profit sharing are paid on an annual or semiannual basis and are tied to company performance.

Bonuses may be used when a department surpasses its goal and that department cannot affect profit. Manufacturing is a good example of this. Manufacturing may perform at superefficient rates, surpassing all previous performance, but business conditions called for reducing product price, resulting in lower profits. Should manufacturing be penalized for doing outstandingly well? Most think not, and the bonus system results. Most bonus systems are tied to specific objectives.

Profit sharing sets a part of the company's profits aside to be divided among eligible employees according to base salary or job grade. Distribution of profits can be made quarterly, annually, or even deferred until retirement. However, the primary disadvantage of bonus and profit-sharing is the lag between effort and reward.

Bonus and profit sharing increase employee morale, reduce turnover, reduce grievances, and promote the feeling of being a part of the company, but neither system is a motivator of increased productivity. The employee sees his or her efforts too far removed from the result. Small companies may find more success with either system than larger companies.

Organized labor dislikes individual incentive systems which promote competition among employees and reward individual productivity. They prefer plans that are universally applied throughout the plant and apply equally for all workers, such as bonuses and profit sharing. Personally, I dislike rewarding nonproductive people in any way.

Sears Roebuck, Lincoln Electric, IBM, Burlington, Xerox, Signode, and Polaroid are among over 33,000 profit-sharing plans in the United States. At a meeting of the Profit Sharing Council of America, Mr. Arthur Wood, then chairman of Sears Roebuck, stated the following:

> Profit sharing is not the first step in building a program of sound employee relations, but the last step. If a company has a good employee selection program . . . if a company provides training for employees so that they become productive on the job quickly, and have an opportunity to prepare themselves for additional responsibility; if a company pays its employees fairly; if a company has developed personnel policies that provide the individual with security against arbitrary acts of supervision; if a company has demonstrated a concern about the morale of its organization . . . if a company provides its employees with an opportunity to participate in benefit programs that offer protection against the hazards of life . . . then that company should consider profit sharing

Profit sharing can be rewarding for employees, and these rewards will return loyalty and higher morale. If profits dwindle, so will loyalty and morale.

Factors Other Than Labor

Employees can and do affect cost based on much more than their labor. Material utilization can affect cost greatly. Machine utilization can cost 100 times more than the operator's cost. Quality or reject rate can affect not only cost but customer satisfaction and repeat business. If an employee can be properly motivated, he or she can save money on each of the aforementioned items. The present cost of material can be calculated, and a savings program can be established to share the savings with the employees. Every dollar given to employees will return another dollar to the company if a 50/50 split is used.

Machines can cost $10,000 per hour operating cost. If we know the history of downtime, we can build a savings program and share with employees. An experience with an oil quart canning line should that the machine had 24,000 cans per hour capacity, yet only 84,000 cans were packed on an average day ($3\frac{1}{2}$ hours). A performance improvement program was developed to pay the employees a 10% bonus for every hour of additional runtime squeezed out of the machine. Forty-five percent was the maximum bonus potential, and if employees accomplished even one hour of additional production, an overhead of $1,000 per hour was saved.

An incentive system can be designed for any cost. What are we waiting for?

QUESTIONS

1. What are the three ways of paying people?
2. What is the difference between:
   a. Salaried and hourly?
   b. Exempt and nonexempt?
   c. Day work and measured day work?
3. How is a job evaluation system used?
4. What are the six steps of a job evaluation study?
5. What will an incentive system do?
6. What are the two basic types of incentives?
7. What did the Mitchell Fein study of incentive systems show?
8. What percentage of the U.S. labor force is on incentive pay systems?
9. Review the seven philosophies of an incentive system.
10. What problems can be anticipated when starting an incentive system?
11. Why are these problems important to the industrial technologist?
12. List four individual incentive systems and describe each.
13. List three group incentive systems and describe each.
14. When are bonuses and profit sharing useful?
15. What factors other than labor can be included in incentive systems?

CHAPTER 14

Mil-Std 1567A

During June of 1975, the United States Air Force published Military Standard 1567A, establishing minimum requirements for mandatory work measurement programs in the private aerospace industries. Any company doing business with the military has to have a work measurement program.

The significant requirements of the military standard are as follows:

1. The plan must cover at least 80% of all direct manufacturing labor (touch labor) with engineered standards.
2. Accuracy must be of at least ±25% at a 95% confidence level with data to prove it.
3. There must be written variance analysis of labor performance reports.
4. There must be an auditing system of coverage and accuracy.

Mil-Std 1567 (USAF) has created a large demand for work measurement people. The military standard has permeated all government work today, and since the government is the largest industrial customer in the world, a knowledge of Mil-Std 1567 (USAF) is necessary for all industrial technologists.

An article entitled "Work Measurements: The Flap Over Mil.Std.1567 (USAF)," which appeared in *Industrial Engineer* magazine in November 1976 (pp. 14–25), is a good history lesson of this military standard and a lesson on change. A great deal of resistance to Mil-Std 1567A was put forth during its first ten years. Much time and effort was lost due to the conflict between government and management.

The profession of industrial engineering has promoted work measurement as a scientific management tool for over forty years. The profession has no problem with Mil-Std 1567A, but management dislikes anyone telling it how to manage. The reporting requirements are extra work and have the force of "law."

My own belief is that Mil-Std 1567A is not strong enough, and too many managers are fighting it instead of making it work for them. The new generation of industrial managers and engineers will hopefully make the work measurement produce savings for all of us taxpayers.

The text of Mil-Std 1567A is included in this chapter for two reasons:

1. The industrial technologist who has read this textbook to this point understands the technical information and agrees that Mil-Std 1567A is not difficult to attain.

2. The government is such a big customer that Mil-Std 1567A will be a part of every motion and time study technician's future.

Thirteen pages of the appendix of Mil-Std 1567A were omitted to save space. A complete text can be obtained from the U.S. Government Printing Office.

JOINT AGREEMENT ON SUPPORT
OF MIL-STD-1567A, WORK MEASUREMENT

1. Interservice cooperation is required to ensure MIL-STD-1567A, Work Measurement is applied widely and consistently.

2. MIL-STD-1567A is an essential weapon in DOD's cost reduction arsenal. When applied effectively, it increases discipline in contractor work measurement systems, thereby improving productivity and efficiency. The Standard does this by requiring contractors to monitor direct labor performance while simultaneously providing Government visibility into their progress. Specific benefits resulting from this emphasis and visibility include:

   a. Objective performance measurement by comparing "actual" hours expended to a "standard" hour baseline.

   b. Methods improvements resulting in increased efficiency and standard hour content reduction.

   c. Early identification of potential quality problems since production problems often lead to quality defects.

   d. Easy comparison of alternative manufacturing methods.

   e. Realistic man-loading based on past performance and planned improvements.

   f. A better foundation for pricing and negotiating.

3. Work measurement systems are "process," not "product," oriented. Therefore, factory-wide systems are the most effective. Since MIL-STD-1567A is applied on a contract-by-contract basis, widespread application maximizes DOD benefits while minimizing system implementation and maintenance costs. It is clear that cooperation and coordination among the Commands is critical.

4. We endorse the consistent contractural application of MIL-STD-1567A on major acquisitions meeting the applicability criteria defined therein. Senior focal points in each Command will be designated. These individuals will:

a. Coordinate Command policy and implementation strategies.

b. Integrate these activities with those of the other Command focal points.

c. Ensure MIL-STD-1567A provisions are applied effectively and expeditiously.

5. Command MIL-STD-1567A focal points are directed to report their progress to the JLC in one year.

RICHARD H. THOMPSON
General, USA
Commander
US Army Materiel Command

S. A. WHITE
Admiral, USN
Chief of Naval Material
Naval Material Command

EARL T. O'LOUGHLIN
General, USAF
Commander
Air Force Logistics Command

LAWRENCE A. SKANTZE
General, USAF
Commander
Air Force Systems Command

DATE: 8 January 1985

MIL-STD-1567A
11 March 1983

FOREWORD

The purpose of this standard is to assist in achieving increased discipline in contractors' work measurement programs with the objective of improved productivity and efficiency in contractor industrial operations. Experience has shown that excess manpower and lost time can be identified, reduced, and continued method improvements made regularly where work measurement programs have been implemented and conscientiously pursued.

Active support of the program by all affected levels of management, based on an appreciation of work measurement and its objectives, is vitally important. Work Measurement and the reporting of labor performance is not considered an end in itself but a means to more effective management. Understanding the implication inherent in the objectives of the work measurement program will promote realization of its full value. It is important that objectives be presented and clearly demonstrated to all personnel who will be closely associated with the program.

The following are benefits which can accrue as a result of the employment of a work measurement program.

(a) Achieving greater output from a given amount of resources.

(b) Obtaining lower unit cost at all levels of production because production is more efficient.

(c) Reducing the amount of waste time in performing operations.

(d) Reducing extra operations and the extra equipment needed to perform these operations.

(e) Encouraging continued attention to methods and process analysis because of the necessity for achieving improved performance.

MIL-STD-1567A

(f) Improving the budgeting process and providing a basis for price estimating, including the development of Government Cost Estimates and should cost analyses.

(g) Acting as a basis for planning for long-term manpower, equipment, and capital requirements.

(h) Improving production control activities and delivery time estimation.

(i) Focusing continual attention on cost reduction and cost control.

(j) Helping in the solution of layout and materials handling problems by providing accurate figures for planning and utilization of such equipment.

(k) Providing an objective and measured base from which management and labor can project piecework requirements, earnings and performance incentives.

Feedback on the success or difficulties encountered (benefits and costs) in the application of this standard on specific contracts is encouraged. Contractor/ industry and Government experience should be forwarded to the address indicated on page ii.

While recognizing the benefits that may normally be expected from the requirement for a work measurement system, it is DOD policy to selectively apply and tailor standardization documents to ensure their cost-effective use in the acquisition process. Each program office should carefully consider, within DOD and Service guidelines, benefits and costs of imposing MIL-STD-1567 on each specific acquisition. Contractors may propose document application and tailoring modifications with supporting rationale for such modifications.

MIL-STD-1567A

The DOD is committed to development and coordination with industry of detailed application guidance to accompany MIL-STD-1567. The purpose of this guidance is to provide noncontractual information on when and how to use the document, the source of and flexibility inherent within specific document requirements, information on what is required to satisfy document requirements, and the extent of Government review and approval. The guidance is intended to promote consistency in application and interpretation of MIL-STD-1567 requirements. Until this guidance can be issued in the form of an ''Application Guidance'' appendix to MIL-STD-1567, or in a separate Military Handbook, the following applies:

(a) Use and correct application of appropriate predetermined time systems can be assumed to satisfy Government requirements for system accuracy.

(b) The contractor and the Government are encouraged to come to an early agreement (possibly in the form of a Memorandum of Understanding) of what constitutes an acceptable system satisfying the intent of this standard.

(c) Care should be exercised in the use of a work measurement system to ensure that the overall intent is not lost. Management understanding and attention to the manufacturing process is necessary for increased productivity. Work measurement provides one of the tools; however, misuse could result in reduced workforce motivation and productivity.

MIL-STD-1567A

CONTENTS

Supersedes page vi of 11 March 1983.

MIL-STD-1567A

Supersedes page vii of 11 March 1983.

MIL-STD-1567A
11 March 1983

MILITARY STANDARD

WORK MEASUREMENT

1. SCOPE

1.1   Purpose. This standard requires the application of a disciplined work measurement program as a management tool to improve productivity on those contracts to which it is applied. It establishes criteria which must be met by the contractor's work measurement programs and provides guidance for implementation of these techniques and their use in assuring cost effective development and production of systems and equipment.

1.2   Applicability. This standard is applicable to new/follow-on contracts, including modifications, as shown in paragraphs 1.2a, 1.2b, 1.2c, and 1.2.1 below. The dollar thresholds indicated are to be based on the current Five Year Defense Program (FYDP) budget submissions.

　　　a. Full-scale acquisition program developments which exceed $100 million.

　　　b. Production, which may include some types of depot level maintenance repair or overhaul, that exceeds $20 million annually or $100 million cumulatively. It shall not be applied to contracts or subcontracts for construction, facilities, off-the-shelf commodities, time and materials, research, study, or developments which are not connected with an acquisition program.

　　　c. This standard is not applicable to ship construction, ship system contracts which have low volume non-repetitive production runs, or service-type contracts.

1.2.1   Subcontracting. When this standard is applied to prime development or production contracts, it shall also be applied to related subcontracts and/or modifications which exceed $5 million annually or $25 million cumulatively. If it is determined by the prime contractor that such application is not cost effective or inappropriate for other reasons, the prime contractor may request the Government to waive the specific application. Requests for waivers shall be supported with the data used to make the determination.

MIL-STD-1567A
11 March 1983

1.3   Contractual Intent. This standard requires the application of a
documented work measurement system. This standard further requires that the
contractor apply procedures to maintain and audit the work measurement system.
It is not the intent of this standard to prescribe or imply organization structure,
management methodology, or the details of implementation procedures.

1.4   Corrective Actions. When surveillance by the contractor or the
Government discloses that the work measurement program does not meet the
requirements of this standard, a plan shall be initiated to expeditiously assure
that corrective measures shall be implemented, demonstrated, and documented.
The contractor's system is subject to disapproval by the Government whenever it
does not meet the requirements of this standard.

1.5   Documentation. The work measurement program shall include sufficient
documentation to assure effective operation of the program and to provide for
internal audits as required by paragraph 5.14. Documentation shall specify
organizational responsibilities, state policies, and provide operational procedures
and instructions. The results of contractor system audits and plans for corrective
actions shall be made readily available to the Government for review.

2. REFERENCED DOCUMENTS.

Not Applicable

3. DEFINITIONS.

3.1   Actual Hours. An amount determined on the basis of time incurred as
distinguished from forecasted time. Includes standard time properly adjusted for
applicable variance.

3.2   Earned Hours. The time in standard hours credited to a worker or group
of workers as the result of successfully completing a given task or group of
tasks: usually calculated by summing the products of applicable standard times
multiplied by the completed work units.

MIL-STD-1567A
11 March 1983

3.3   Labor Efficiency. The ratio of earned hours to actual hours spent on same increments of work during.a reporting period. When earned hours equal actual hours, the efficiency equals 100%.

3.4   Methods Engineering. The analyses and design of work methods and systems, including technological selection of operations or processes, specification of equipment type, and location.

3.5   Operation Analysis. A study which encompases all those procedures concerned with the design or improvement of production, the purpose of the operation or other operations, inspection requirements, materials used and the manner of handling material, setup, tool equipment, working conditions, and methods used.

3.6   Predetermined Time System. An organized body of information, procedures, and techniques employed in the study and evaluation of manual work elements. The system is expressed in terms of the motions used, their general and specific nature, the conditions under which they occur, and their previously determined performance times.

3.7   Realization Factor.

    (a) A ratio of total actual labor hours to the standard earned hours.

    (b) A factor by which labor standards are multiplied when developing actual/projected manhour requirements.

3.8   Subcontract. A contract between the prime contractor and a third party to produce parts, components, or assemblies in accordance with the prime contractor's designs, specifications or directions and applicable only to the prime contract.

3.9   Touch Labor. Production labor which can be reasonably and consistently related directly to a unit of work being manufactured, processed, or tested. It involves work affecting the composition, condition, or production or a product; it may also be referred to as ''hands-on labor'' or ''factory labor.''

MIL-STD-1567A
11 March 1983

NOTE: As used in this standard, touch labor includes such functions as machining, welding, fabricating, setup, cleaning, painting, assembling, functional testing of production articles, and that labor required to complete the manually-controlled process portion of the work cycle.

3.10   Touch Labor Standard. A standard time set on a touch labor operation.

3.11   Type I Engineered Labor Standards. These are standards established using a recognized technique such as time study, standard data, a recognized predetermined time system or a combination thereof to derive at least 90% of the normal time associated with the labor effort covered by the standard and meeting requirements of paragraph 5.1. Work sampling may be used to supplement or as a check on other more definitive techniques.

3.12   Type II Labor Standard. All labor standards not meeting the criteria established in paragraph 5.1.

3.13   Standard Time Data. A compilation of all elements that are used for performing a given class of work with normal elemental time values for each element. The data are used as a basis for determining time standards on work similar to that from which the data were determined.

3.14   Touch Labor Normal/Standard Time. Normal time is the time required by a qualified worker, to perform a task at a normal pace, to complete an element, cycle or operation, using a prescribed method. The personal, fatigue, and unavoidable delay allowance added to this normal time results in the standard time.

3.15   Operation. (1) A job or task consisting of one or more work elements, normally done essentially in one location; (2) The lowest level grouping of elemental times at which P F & D allowances are applied.

3.16   Element. A subdivision of the operation composed of a sequence of one or several basic motions and/or machine or process activities which is distinct, describable, and measurable.

MIL-STD-1567A
11 March 1983

4. <u>GENERAL REQUIREMENTS</u>.

4.1   <u>General</u>. Minimum requirements which must be met in the implementation of an acceptable work measurement program are:

    a. An explicit definition of standard time that shall apply throughout the jurisdiction of work measurement.

    b. A work measurement plan and supporting procedures.

    c. A clear designation of the organization and personnel responsible for the execution of the system.

    d. A plan to establish and maintain engineered labor standards to known accuracy.

    e. A plan to conduct methods engineering studies to improve operations and to upgrade Type II labor standards to Type I engineered Labor Standards in accordance with requirement of paragraph 5.4.

    f. A defined plan for the use of labor standards as an input to budgeting, estimating, production planning, and "touch labor" performance evaluation.

    g. A plan to ensure that system data is corrected when labor standards are revised according to paragraph 5.11 below.

5. <u>SPECIFIC REQUIREMENTS</u>.

5.1   <u>Type I Engineered Labor Standards</u>. All Type I standards must reflect an accuracy of ± 10% with a 90% or greater confidence at the operation level. For short operations, the accuracy requirement may be better met by accumulating small operations into super operations whose times are approximately one-half hour. Type I standards must include:

    a. Documentation of an operations analysis.

MIL-STD-1567A
11 March 1983

   b. A record of standard practice or method followed when the standard was developed.

   c. A record of rating or leveling.

   d. A record of the standard time computation including allowances.

   e. A record of observed or predetermined time system time values used in determining the final standard time.

5.1.1   Predetermined Time Systems. It is not the intent of this Military Standard to challenge the accuracy of those predetermined time systems whose inherent accuracy meets the requirements of paragraph 5.1. However, when a predetermined time system is used, it shall be incumbent on the contractor to demonstrate to the Government that the accuracy of the original data base has not been compromised in application or standards development.

5.2   Operations Analysis. Operations analysis is considered an integral part of the development of a Type I Engineered Labor Standard. An operations analysis shall be accomplished and recorded prior to the determination of a Type I standard; and in the improvement of established labor standards.

5.3   Standard Data. The contractor shall take full advantage of available standard time data of known accuracy and traceability.

5.4   Labor Standards Coverage. The contractor shall develop and implement a Work Measurement Coverage Plan which provides a time-phased schedule for achieving 80% coverage of all categories of touch labor hours with Type I standards. (See 3.9, Touch Labor)

5.4.1   Cost Trade-off Analysis. The Work Measurement Coverage Plan shall be based on cost trade-off analyses which consider the status and effectiveness of the contractor's existing work measurement program.

MIL-STD-1567A
11 March 1983

5.4.2 Initial Coverage. Type II Standards are acceptable for initial coverage. All Type II standards shall be approved by the organization(s) responsible for establishing and implementing work measurement standards and estimating when Type I Standards have not yet been developed.

5.4.3 Upgrading. The Work Measurement Touch Labor Coverage Plan shall provide a schedule for upgrading Type II to Type I Standards.

5.5   Leveling/Performance Rating. All time studies shall be rated using recognized techniques.

5.6   Allowances. Allowances for personal, fatigue, and unavoidable delays shall be developed and included as part of the labor standard. Allowances should not be excessive or inconsistent with those normally allowed for like work and conditions.

5.7   Estimating. The contractor's procedures shall describe how touch labor standards are utilized to develop price proposals.

5.8   Use of Labor Standards. Labor standards shall be used:

5.8.1 Budgets, Plans, and Schedules. As an input to developing budgets, plans and schedules, when available.

5.8.2 Touch Labor Hours. As a basis for estimating touch labor hours when issuing changes to contracts and as a basis for estimating the prices of initial spares, replenishment spares and follow-on production buys, when available.

5.8.3 Measuring Performance. As a basis for measuring touch labor performance.

5.9   Realization Factor. When labor standards have been modified by realization factors, major elements which contribute to the total factor shall be identified. The analysis supporting each element shall be available to the Government for review.

MIL-STD-1567A
11 March 1983

5.10   Labor Efficiency. A forecast of anticipated touch labor efficiency shall be used in manpower planning, both on a long-range and current scheduling basis.

5.11   Revisions. Labor standards shall be reviewed for accuracy and appropriate system data revision made when changes occur to:

     a. Methods or procedures

     b. Tools, jigs, and fixtures

     c. Work place and work layout

     d. Specified materials

     e. Work content of the job

5.12   Production Count. Work units shall be clearly and discretely defined so as to cause accurate measurement of the work completed and shall be expressed in terms of completed:

     a. End items

     b. Operations

     c. Lots or batches of end items

5.12.1 Partial Credit. In those cases where partial production credit is appropriate, the work measurement procedures shall define the method to be used to permit a timely and current production measure.

5.13   Labor Performance Reporting. The contractor's work measurement program shall provide for periodic reporting of labor performance. The report shall be prepared at least weekly for each work center and be summarized at each appropriate management level; it shall indicate labor efficiency and compare current results with pre-established contractor goals.

MIL-STD-1567A
11 March 1983

5.13.1 Variance Analysis. Labor performance reports shall be reviewed by supervisory and staff support functions. When a significant departure from projected performance goals occurs, a formal written analysis which addresses causes and corrective actions shall be prepared.

5.13.2 Report Retention. Performance reports and related variance trend analyses shall be retained for a six-month period.

5.14 System Audit. The contractor shall use an internal review process to monitor the work measurement system. This process shall be so designed that weaknesses or failures of the system are identified and brought to the attention of management to enable timely corrective action. Written procedures shall describe the audit techniques to be used in evaluating system compliance.

5.14.1 Scope of Audit. The audit shall cover compliance with the requirements of this standard at least annually. The audit, based upon a representative sample of all active labor standards and work measurement activities, shall determine:

a. The validity of the prescribed method and the accuracy of the labor standard time values as validated against the data baseline.

b. Percent of coverage by Type I and Type II labor standards.

c. Effectiveness of the use of labor standards for planning, estimating, budgeting, and scheduling.

d. The timeliness, accuracy, and traceability of production count reporting.

e. The accuracy of labor performance reports.

f. The reasonableness and attainment of efficiency goals established.

g. The effectiveness of corrective actions resulting from variance analyses.

5.14.2 Audit Reports. A copy of the audit finding shall be retained in company files for at least a two-year period and shall be made available to the Government designated representative for review upon request.

MIL-STD-1567A

APPENDIX

3.   DEFINITIONS

3.1   Actual Hours.

A.   The unadjusted (actual) time charged by an operator, or group of
operators, to accomplish an operation or task as covered by a Type I or Type II
labor standard (i.e., measured work). It might not include time charges for
"unmeasured" work (operations or tasks without labor time standards). It might
not include actual time charges for work beyond the control of the operator(s)
such as "idle" or "lost" time due to delays awaiting material, parts, or
inspection, or machine downtime. Also, it might not include rework/repair/scrap
due to engineering changes or vendor problems. It normally does include
opeartor(s) time charges associated with rework/repair/scrap due to operator
error.

3.2   Earned Hours.

A.   Self-explanatory.

3.3   Labor Efficiency

A.   Labor Efficiency $\% = \dfrac{\text{Earned Hours}}{\text{Actual Hours}} \times 100$

This is a measure of "operator" efficiency against a particular task or
aggregation of tasks. Labor efficiency is not necessarily the inverse of
"Realization Factor" (para 3.7). Realization Factor generally measures overall
performance ("shop, product line, plant").

3.4   Methods Engineering.

A.   Methods engineering is the function of selecting or adapting the most
cost-effective process technologies to fulfill a given manufacturing (including
quality) requirement. Generally, methods engineering studies should include a
clear description of the work station/layout, the method to be employed, and the
work unit under consideration. The method chosen serves as the basis for
development of the touch labor standard (see para 3.10).

MIL-STD-1567A
APPENDIX

B.   Methods engineering also includes those functions which attempt to improve existing processes or reduce existing work content. The result of such methods engineering is often called methods improvement.

C.   For some contractors, development, adaptation, or design of new processes may be accomplished by functions other than industrial engineering. Selection of existing processes may also be accomplished by functions other than industrial engineering.

D.   Methods engineering functions can be identified on an individual basis and may include the proper use or application of specific operations or processes determined by diverse functional organizations.

E.   Although other efforts, such as Suggestion Program, Zero Defects, Value Engineering, etc., provide a positive impact on methods, they are normally not an integral part of a contractor's methods engineering program/system. These efforts of and by themselves do not satisfy the methods engineering requirement.

3.5   Operation Analysis.

A.   An operation analysis is a technique used to evaluate all productive and nonproductive elements of an operation with an emphasis on productivity improvement. The analysis could consider such factors as potential for design changes, production rate, and labor content. Each analysis should emphasize those areas which are most cost-effective.

B.   Operations analysis should contain information for as many of the following areas as possible: (1) A description of the tools, workplace, and any other physical conditions involved in the work; (2) A detailed description of what work is being done (including tasks that ensure product quality) and why it is being accomplished in this manner (i.e., balancing cost with results); (3) A chronological sequencing of when events take place in the work cycle; (4) A description of how the worker applies himself to the tools, equipment or parts to accomplish the job; (5) The constraints on where the work is being done; (6) A description of who is doing the job (skills required, assembly, mechanic, helper, machinist, etc.).

MIL-STD-1567A
APPENDIX

C.   The above analysis can be made using many techniques available to the trained labor standards engineer. These procedures will reveal to the analyst whether or not the optimum working conditions exist prior to developing the standard or standard data. It will also record existing conditions when the standard is developed. It will result in a record of all physical aspects of the task for audits and for updates when any working conditions change in the work place.

3.6   Predetermined Time System.

A.   Self-explanatory.

3.7   Realization Factor.

A.   Realization Factor $= \dfrac{\text{Total Actual Hours}}{\text{Earned Hours}}$

Where "Total Actual Hours" include all manufacturing labor hours (reconcilable with payroll hours) associated with tasks represented by the "Earned Hours" in the denominator, including "lost time" or "idle time" accounts and/or "off standard" or "unmeasured" work (application guide paras 3.1.A, 3.14.C and 4.1.b.B). Realization Factor is generally a measure of overall performance ("shop, product line, plant"). Realization factor is not necessarily the inverse of "Labor Efficiency" (para 3.3). Labor Efficiency is used to measure "operator" performance against a particular task or aggregation of tasks.

4.1.b

A.   The work measurement plan is the written documentation which describes the contractor's work measurement system. The plan should describe how the contractor will comply with the General (para 4) and Specific (para 5) requirements of the Standard. If a review by the government or the contractor determines that the system does not comply with the requirements of MIL-STD-1567A (or such compliance as is contractually required), the plan will normally also include a time-phased milestone schedule to accomplish full compliance. The time-phased milestone schedule should recognize the status of the contractor's existing work measurement system and be mutually agreed to by the buying activity and the contractor.

MIL-STD-1567A
APPENDIX

B.   The work measurement plan should also discuss how ''lost time'' or ''idle time'' accounts (application guide paras 3.1.A and 3.14.C) and/or ''off standard'' or ''unmeasured'' work will be monitored, evaluated, and reported. ''Off standard'' or ''unmeasured'' work is work which must be accomplished by touch labor personnel to support a particular task but which is not directly charged against that task (application guide para 3.1.A).

C.   The work measurement plan should be consistent with the contractor's disclosure statement.

4.1.c

A.   Self-explanatory.

4.1.d

A.   Self-explanatory.

4.1.e

A.   Methods improvements to improve operations and reduce work content is an integral part of a good work measurement system. An aggressive methods improvement program is of particular interest to the government because effective methods improvement can simultaneously reduce ''variance'' and standard hour work content. Periodically setting and achieving time-phased goals for significant standard hour content reduction are good indications of a disciplined and effective methods improvement program.

B.   Normally, operations identified by variance reports or other sources will be cost-effectively considered and selected for methods improvement studies. Candidate operations could include those having a relatively high actual labor content, those with a history of shop floor discrepancies (including poor quality or production bottlenecks), or those independently identified by industrial engineers or manufacturing engineers as good candidates for significant standard hour content reduction.

MIL-STD-1567A
APPENDIX

C.   All methods improvement studies which affect the touch labor standard should be documented and made available to the government for review upon request.

4.1.f

A.   Self-explanatory.

4.1.g

A.   Self-explanatory.

5.   SPECIFIC REQUIREMENTS

5.1   Type I Engineered Labor Standards.

A.   To demonstrate compliance to this paragraph, all standard data (plus a record of revisions and audits pertaining to the data) should be retained by the organization which developed the data. Specific documentation requirements should be limited to those necessary for the contractor or government to have reasonable confidence that recognized industrial engineering techniques have been correctly applied in a consistent and accurate manner. If supporting documentation is not available for systems established prior to the application of this standard, a statistically valid sampling approach may be used to attempt to ''reconstruct'' the labor standards. If such reconstruction confirms the accuracy of existing labor standards, no additional documentation will be required for those labor standards from which the sample was selected.

B.   As used in the Standard, ''accuracy'' is intended to mean the degree of correctness or exactness of labor standard operations. Normally, the degree of accuracy is determined by comparing the average time of a statistically valid sample of measurements with the proposed labor standard. Proper use and application of appropriate predetermined time systems can be assumed to satisfy Government requirements for system accuracy.

5.1.a

MIL-STD-1567A
APPENDIX

A.   The contractor should be prepared to provide evidence that cost-effective operations analysis was performed. Video tapes or computer-aided-design/ computer-aided-manufacturing (CAD/CAM) displays may be used to supplement or replace documentation as appropriate (see application guide para 3.5).

5.1.b

A.   Two ways (others may also be appropriate) in which standard practice or method could be demonstrated are:

    1. A separate method sheet could be provided which includes all elements of the operation to be performed, in sequence, with all tooling and material requirements identified.

    2. An elemental breakdown could be provided adjacent to the cycle times. With predetermined time systems, the elements may be further broken down into basic movements opposite the resulting times.

B.   Introduction of new technology or new or modified equipment will almost always impact the original labor time standard. When this happens, elements of the time standard should be revised or deleted. Revisions to the original labor time standard should be traceable to, and a part of, standard data documentation.

5.1.c

A.   A predetermined time system assigns a pre-established time value (which does not permit interpretation by analysts) to each basic motion encountered. Rating or leveling of this data is not necessary.

B.   If time studies are used to develop labor standards, an appropriate performance rating system should be used. The person doing the rating should be skilled and well-trained in the application of whatever system is utilized. Training devices such as films or video tapes should be used to improve proficiency. The contractor should record and retain a schedule of training and the results of practice ratings for each engineer who uses time studies to develop labor standards. The observer should record the performance of the operator studied in order to determine how the allowed time was calculated.

MIL-STD-1567A
APPENDIX

5.1.d

A.   The computations necessary to develop the standard should be included
with the back-up data for developing a standard. These should be available for
audit purposes by contractor and by the government.

5.1.e

A.   Self-explanatory.

5.1.1   Predetermined Time Systems.

A.   A standard data package with its supporting information should contain
concise, complete data on the conditions and method employed. Demonstration
of standard data development is not difficult if an audit trail is included in the
package. Check studies in the standrd data (to determine if all elements have
been addressed) should also normally be a part of each package. Check studies
are normally a part of the contractor's internal audit verifying accuracy and (if
available) should be provided to the government for review, upon request.

5.8.1   Budgets, Plans, and Schedules.

A.   Self-explanatory.

5.8.2   Touch Labor Hours.

A.   As stated in application guide para 5.7, the use of labor standards in
estimating is an area of emphasis to the government. The government intends to
require labor standards to be used as the basis for establishing touch labor hours
for changes to contracts, initial and replenishment spares, and follow-on
production buys, when such standards are available.

B.   Since the design and/or manufacturing processes may not be stable during
development or prior to initial production, touch labor standards may not be
available for estimates associated with such efforts. This is why MIL-STD-
1567A specifies labor standards are to be used for estimating "...follow-on
production buys, when available."

MIL-STD-1567A
APPENDIX

C.   Labor standards may be adjusted by a realization factor to arrive at a projected unit value. (See application guide para 5.9.C for a discussion of realization factor elements to be identified and analyzed.) Appropriate improvement curves may be selected and applied. The realization factor and corresponding improvement curve should be selected giving due consideration to appropriate factors such as the program environment (including design stability), the configuration baseline, past inefficiencies which have been, or should be, corrected, anticipated methods and process changes (including expected performance increases due to methods improvement and variance reduction), concurrency of design and production, and working conditions.

5.8.3   Measuring Performance.

A.   Performance measurement is another area of emphasis to the government. Consequently, touch labor performance should be measured at the work center/ department level and summarized and reported at higher levels meaningful to a company's organizational structure. This reporting summary should complement monitoring, analyzing, and improving performance in accordance with pre-established contractor improvement goals.

B.   Although work measurement systems are generally process—not product—oriented, the government is primarily interested in cost-effectively acquiring end items (a "product" whether it be a spare part, component, or system). Therefore, contractors should summarize performance to the end item (or other selected production count work unit) level when requested to do so by the government. Such performance is usually summarized as labor hours (total actual, actual or earned standard) per "equivalent" unit; or per completed unit. If "equivalent units" are used, they should be calculated as described in para 5.12.1.A of this guide. The use of equivalent units minimizes the impact of "lag" time between manufacture of piece parts, subsystems, or components, and completion/acceptance of end items. However, unless "fabrication" and "assembly" tasks are proceeding proportionately, equivalent unit reporting could "skew" performance indications and present a misleading picture of current performance. This is because fabrication operations generally experience a higher labor efficiency (para 3.3) and a lower realization factor (para 3.7) than assembly operations. If there is a temporary preponderance of either fabrication or assembly work during a particular reporting period, performance indicators may be "biased" toward either fabrication or assembly.

MIL-STD-1567A
APPENDIX

C.   Performance should be measured as either labor efficiency (para 3.3) or realization factor (para 3.7).

D.   If labor efficiency is chosen as the performance measurement unit, "lost time" or "idle time" accounts as discussed in application guide para 3.14.C and/or "off standard" or "unmeasured" work should be monitored, evaluated, and reported separately in accordance with the contractor's work measurement plan and supporting procedures (application guide para 4.1.b.B). (See also application guide para 3.1.A).

5.9   Realization Factor.

A.   Elements of the realization factor should be identified and quantified in sufficient detail to permit a reasonable "...analysis supporting each element..."

The word "identified" is intended to imply more than a "listing" of realization factor elements. If the elements of the realization factor were simply "listed," the required "...analysis supporting element..." would not be possible.

B.   This is also an area of government emphasis since realization factors must be described in sufficient detail to permit the government to use work measurement data (for example, labor standards modified by realization factors) in a should cost approach to pricing and negotiating contracts.

C.   Typical elements of realization could include "learning" (such as familiarization and instruction in reading engineering drawings and operation sheets, and in using the appropriate method); "technical" (such items as engineering changes, design errors, fit problems, operation sheet errors, tooling errors, sequencing errors, manufacturing/design engineering coordination, and scrap/rework/repair/re-inspection); "logistics" (such items as incorrect hardware, part shortages, and waiting for inspection); and "miscellaneous" items such as excessive overtime and/or fatigue (beyond that included in the basic touch labor standard). The significance of individual elements often varies depending on program environment characteristics such as design stability, production process maturity, and operation complexity. The elements should be identified and analyzed at an organizational level consistent with the cost effective gathering of supporting data while also keeping in mind the impact of realization factors in performance measurement, and planning, budgeting, scheduling, and estimating.

MIL-STD-1567A
APPENDIX

D.   Setting and achieving aggressive goals for reducing realization factors beyond historical "improvement curve" effects will be a prime factor in government review of contractor performance. Achieving aggressive performance goals is an excellent indicator of contractor compliance to MIL-STD-1567A provisions.

5.10   Labor Efficiency.

A.   Setting and achieving aggressive labor efficiency improvement goals (beyond historical "improvement curve" effects) will be viewed by the government as an excellent indicator of contractor compliance to MIL-STD-1567A provisions.

5.11   Revisions.

A.   Other circumstances (for example, worker or supervisor statements) may also indicate labor standards should be reviewed.

5.12   Production Count.

A.   Work unit production count during a specified period is of importance primarily for purposes of monitoring, evaluating, and forecasting contractor performance. Therefore, the contractor should make available to the government (upon request) the count of the total and/or equivalent number of end items completed during a specific time period.

B.   Cost effectiveness, and consistent and accurate results, should be considered when selecting the production count work unit.

5.12.1 Partial Credit.

A.   One method of determining partial credit could be to estimate the "equivalent work units" produced. This could be calculated by dividing the "Earned Hours" (para 3.2) during a specified period for a selected work unit by the total "Touch Labor Standards" (para 3.10) associated with that same work unit. Other methods of determining partial credit for work-in-process may also be appropriate.

MIL-STD-1567A
APPENDIX

5.13   Labor Performance Reporting.

A.   Performance reporting should include labor efficiency (para 3.3) or
realization factor (para 3.7). If labor efficiency is chosen as the unit with which
to report performance, "lost time" or "idle time" accounts and/or any "off
standard" or "unmeasured" work (application guide paras 3.1.A, 3.14.C,
4.1.b.B and 5.8.3.D) not included in "labor efficiency" should also be reported.

B.   Aggressive performance improvement goals (both for standard hour
reduction and for variance improvements) should be established at organizational
levels consistent with those levels at which performance is measured (application
work measurement system which has been previously accepted by the
government, evidence of such may be submitted to demonstrate compliance for
the contract in question. If the contractor does not have a work mesurement
system, the proposal will describe how and when the requirements of DOD
MIL-STD-1567A will be met.

B.   Sample Contract Language for Full Scale Development and Production
Contracts. Insert the following (see related sections 5.4.C and 5.4.D of this
guidance):

Work Measurement Systems:

        (a) The contractor shall establish, maintain, and use in the performance of
his contract, a work measurement system meeting the criteria of DOD MIL-
STD-1567A, Work Measurement. If the contractor does not have a previously
accepted work measurement system, compliance to the requirements of DOD
MIL-STD-1567A as represented in the contractor's phasing plan will be mutually
agreed upon between the contractor and the buying activity. As part of the
acceptance procedure, the contractor shall make available to the government a
description of the work measurement system applicable to this contract in such
form and detail as indicated in DOD MIL-STD-1567A or as mutually agreed to
by the government and the contractor. The audit of the contractor's Work
Measurement System will assure compliance to the requirements of DOD MIL-
STD-1567A.

MIL-STD-1567A
APPENDIX

(b) The contractor shall incorporate DOD MIL-STD-1567A in each subcontract which meets the criteria set for in DOD MIL-STD-1567A. The contractor shall incorporate in the subcontract adequate provisions for demonstrations, review, acceptance, and surveillance of the subcontractor's system. The assessment for subcontractor compliance to the requirements of DOD MIL-STD-1567A will be the responsibility of the contractor unless otherwise mutually agreed to between the government and the contractor. Documented evidence of compliance by the subcontractor will be made available to the government upon request.

(c) If the contractor or subcontractor is operating a work measurement system that has been previously accepted, evidence of such may be submitted in lieu of demonstration and review described above.

(d) Maintenance of a work measurement system in compliance with DOD MIL-STD-1567A constitutes a ''material requirement of the contract'' within the meaning of paragraph (c) (1) of the FAR 52.232-16 Progress Payments Clauses. The parties must agree that, for progress payments clause administration purposes, a predetermined percent of the contract value for maintaining an acceptable work mesurement system will be negotiated.

CHAPTER 15

Assembly Line Balancing

PURPOSE

The assembly line balancing technique is a use of elemental time standards. Its purposes are to

1. Equalize the work load among the assemblers.
2. Identify the bottleneck operation.
3. Establish the speed of the assembly line.
4. Determine the number of work stations.
5. Determine the labor cost of assembly and packout.
6. Establish the percentage work load of each operator.
7. Assist in plant layout.
8. Reduce production cost.

INFORMATION NECESSARY

The assembly line balancing technique must build on previously determined facts:

1. Blueprints and bill of materials from product engineering—what needs to be done
2. Production volume required (schedule) from marketing and production control—which gives us the plant rate. Also known as the R (rate) value of the plant.
3. Elemental time standards from Industrial Engineering—How long each part takes to assemble.

THE PLANT RATE

The plant rate (R value) tells the industrial technologist how fast the assembly line has to run. Every other machine and operation in the plant is keyed to the assembly line—parts must be supplied. The R value is in decimal minutes. An R value of .250 minutes means that a finished product must come off the assembly line every .250 minutes or the plant will not produce enough product. Every other machine and operation in the plant must produce a part every .250 minutes (4 parts/min.), or it will be behind schedule. If two parts are needed per assembly, the R value for that part would be .125 minutes. If a work station or machine was going to take .4 minutes, and the R value was .250, how many machines would you need?

$$\frac{\text{Time standard}}{R \text{ value}} = \frac{.400}{.250} = 1.6 \text{ stations}$$

If this was a fabrication machine, we would add all these fractions of work stations together, and then round up and buy or build that many work stations. However, on an assembly line, we would have to round up immediately and provide two work stations. Each work station would only be 80% loaded (busy), but because the persons at the work station before and after this work station are producing a part every quarter of a minute, these two 80% loaded people will need to stay at their work stations all the time.

PLANT RATE CALCULATION

Assembly line balancing starts with the plant rate calculation. Some information is required from other sources before beginning:

1. The production volume (for example, 1,500/shift). Sales or marketing management determines how many units the company can sell, and the production inventory control department calculates how fast to produce. Factors like seasonality, warehousing costs, training costs, and manufacturing costs enter the production volume determination. The industrial technologist cannot do anything in the way of plant layout or assembly line balancing without a production volume estimate, and the industrial technologist is not a good source of sales information.

2. The allowances expected for the plant. Ten percent allowances have been used throughout this book and will suffice for our example. The plant will be down 48 minutes per day or 10% of the time. We cannot expect to produce products every minute per day. And when the assembly line is stopped for one operator, all operators are stopped.

3. The efficiency must be anticipated. Experience has shown that this plant produces at 75% for the first year of production, when normally 85% could be expected. If the industrial technologist designs a plant to produce at the rate of 100%, what is the

chance of success? The chances are very low, and the technologist had better find a new job. Mathematically, the plant rate calculations are as follows:

$$
\begin{array}{rl}
480 & \text{minutes per shift} \\
-48 & \text{minutes downtime} \\
\hline
432 & \text{minutes available} \\
@\ 75\% & \text{(anticipated first-year performance)} \\
\hline
324 & \text{effective minutes per shift} \\
\div\ 1{,}500 & \text{units per shift needed} \\
=\ .216 & \text{minutes/unit plant rate} \\
4.63 & \text{units every minute}
\end{array}
$$

(4.63×432 minutes/shift divided by $75\% = 1{,}500$ units per shift).

In this example, 4.63 parts each minute must come off this assembly line and every other machine in the plant, or we will not make our goal of 1,500 units per day.

R value is our starting point for assembly line balancing.

STANDARD ELEMENTAL TIME

Time standards for each part must be calculated before parts can be combined into jobs. When designing a new production line, these times are calculated using PTSS or standard data. With an R value of .216, the technologist will combine these elements of work together into jobs. These jobs will have times in multiples of .216 minutes—.216, .432, .648, and the highest .864. Four people doing the same job is normally the largest group because of line layout problems produced with more than four operators.

A step-by-step procedure using the line balance example shown in Figure 15-1 and the form with circled numbers (Figure 15-2) will help you understand the logic and math for solving assembly line balancing problems.

This assembly line balance solution can be greatly improved, and later in this chapter we include a better solution.

FIGURE 15-1 Assembly line balance example

⑦ Operation No.	⑧ Operation Description	⑨ R	⑩ Cycle Time	⑪ No. Stations	⑫ % Load	⑬ Ave. Cycle Time	⑭ Hours/ 1,000	⑮ Pieces/ Hour
5	Assemble	.216	.357	2	.179	85%	7.770	129
10	Assemble	.216	.441	3	.147	70%	11.655	86
15	Cement	.216	.210	1	.210	100%	3.885	257
20	Rivet	.216	.344	2	.172	82%	7.770	129
25	Form carton	.216	.166	1	.166	79%	3.885	257
30	Label	.216	.126	1	.126	60%	3.885	257
35	Packout	.216	.336	2	.168	80%	7.770	129
			1.980	12			46.620	

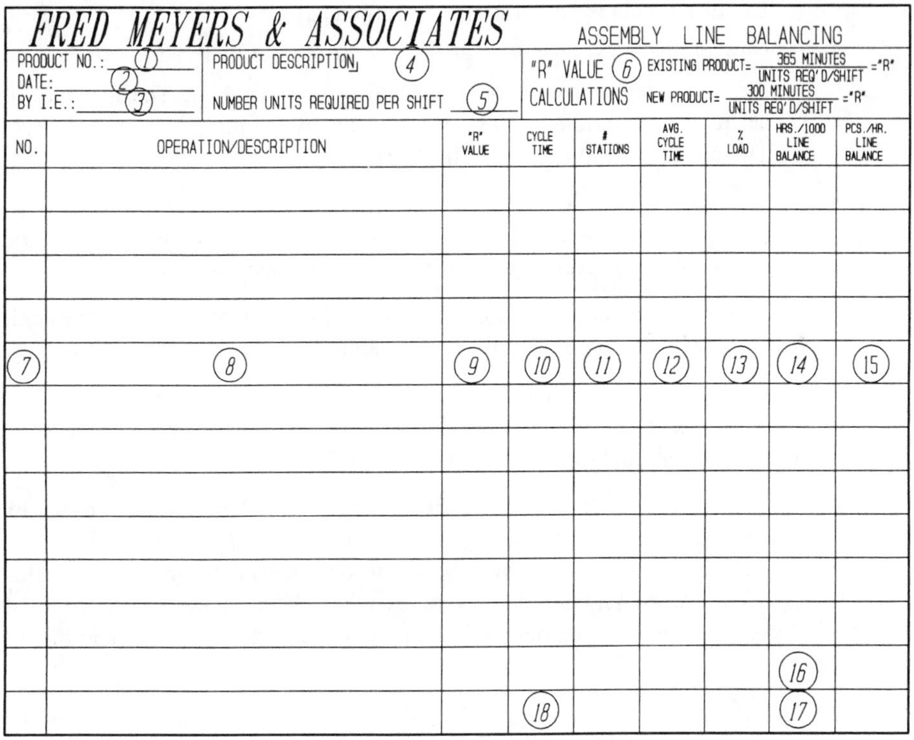
FRED MEYERS & ASSOCIATES — ASSEMBLY LINE BALANCING

PRODUCT NO.: ①
DATE: ②
BY I.E.: ③
PRODUCT DESCRIPTION ④
NUMBER UNITS REQUIRED PER SHIFT ⑤
"R" VALUE ⑥ CALCULATIONS
EXISTING PRODUCT= 365 MINUTES / UNITS REQ'D/SHIFT ="R"
NEW PRODUCT= 300 MINUTES / UNITS REQ'D/SHIFT ="R"

NO.	OPERATION/DESCRIPTION	"R" VALUE	CYCLE TIME	# STATIONS	AVG. CYCLE TIME	% LOAD	HRS./1000 LINE BALANCE	PCS./HR. LINE BALANCE
⑦	⑧	⑨	⑩	⑪	⑫	⑬	⑭	⑮
							⑯	
			⑱				⑰	

FIGURE 15-2 Assembly line balancing form: The step-by-step form

STEP-BY-STEP PROCEDURE FOR COMPLETING THE ASSEMBLY LINE BALANCING FORM

Refer to the assembly line balancing form shown in Figure 15-2.

① Product no.: The product drawing or product part number goes here.

② Date: The complete date of development of this solution.

③ By I.E.: The name of the technician doing the assembly line balance—your name.

④ Product description: The name of the product being assembled.

⑤ Number units required per shift: This is the quantity of production required per shift—given to the technologist by the sales department. The technologist's objective is to get as close to this quantity as possible without going below.

⑥ *R* value: The plant rate has been discussed previously in this chapter, but this block is designed for a specific plant with the following past experience:

a. Existing products have run at 85% efficiency.

b. New products average 70% efficiency during the first year.

c. Eleven percent allowances are added to each standard. The R value in this plant is calculated by dividing 300 or 365 minutes by the number of units per shift (step ⑤). The result is the plant rate—the R value.

⑦ No.: This is a sequential operation number. The use of operation numbers is to give a simple, useful method of referring to a specific job.

⑧ Operation Description: A few well-chosen words can communicate what is being done at this work station. Parts' names and job functions are the key words.

⑨ R value: The R value calculated in block ⑥ goes to the right of each operation. The plant rate is the goal of each work station, and putting the R value on each line keeps that goal clearly in focus.

⑩ Cycle time: The cycle time is the normal time standard set by combining elements of work together into jobs. Our goal is the R value, but that specific number can seldom be achieved. Cycle time can be changed by moving an element of work from one job to another, but elements of work are a large proportion of most jobs. Faster equipment or smarter methods may reduce the cycle time, and this is a good cost reduction tool.

⑪ No. Stations: The number of stations is calculated by dividing the R value ⑨ into the cycle time ⑩ and rounding up. If the number of stations is rounded down, the goal (number of units per shift, ⑤) will not be achieved. Management may round down the number of work stations because of cost, but if they do, they know the goal will not be achieved without overtime, etc. But that is management's decision, not the technologist's. If the number of work stations is rounded down, that work station will be the bottleneck, the restriction, the slowest station, or the 100% station.

⑫ Ave cycle time: The average cycle time is calculated by dividing the cycle time ⑩ by the number of work stations ⑪. This is the speed at which this work station produces parts. If the cycle time of a job is one minute and four machines are required, the average cycle time is .250 minutes ($1.000 \div 4 = .250$) or a part would come out of those four machines every .250 minutes. The best line balance would be for every station to have the same average cycle time, but this never happens. A more realistic goal is to work at getting them as close as possible. The average cycle time will be used to determine the percentage work load of each work station, the next step.

⑬ % Load: The percentage load tells how busy each work station is compared to busiest work station. The highest number in the average cycle time column ⑫ is the busiest work station and is therefore called the 100% station. One hundred percent is written in the % load column. Every other station is compared to this 100% station by dividing the 100% average station time into every other average station time and multiplying the result by 100, equaling the % load of each station. The % load is an indication of where more work is needed or where cost reduction efforts will be most fruitful. If the 100% station can be reduced by 1%, then we will save 1% for every work station on the line.

Example % load calculation:

Look again at the example in Figure 15-1. The average cycle times were .179, .147, .210, .172, .166, .126, and .168. Reviewing these average cycle times reveals that .210 is the largest number and is designated the 100% work station. A good practice is to circle the .210 and the 100% to remind ourselves that this is the most important work station on the line, and no other time standard has any further meaning. Now that the 100% station is determined, the percentage load of every other work station is determined by dividing .210 into every other average cycle time:

$$\text{Operation } 05 = .179 \div .210 = 85\%$$
$$10 = .147 \div .210 = 70\%$$
$$20 = .172 \div .210 = 82\%$$
$$\text{Etc.}$$

Where will the supervisor put the fastest person? Operation 15. Where will the industrial technologist look for improvement or cost reduction? Operation 15, the 100% loaded station.

A good line balance would have all work stations in the 90 to 100% range. One work station below 90% can be used for absenteeism. A new person can be put on this station without holding up the whole line.

⑭ Hours/1,000: The hours per 1,000 units produced can most easily be calculated by multiplying the 100% average cycle time (which is circled on the line balance) by 18.5 (the standard developed in chapter 9 under allowances—see p. 155). 18.5 is the hours per 1,000 for a 1-minute job. 18.5 adds in a constant 10% allowance, and because all normal time standards are in minutes, the number of minutes times 18.5 hours/1,000/minute equals the hours per 1,000 for that job.

From our example, Figure 15-1, .210 was the 100% station:

$$.210 \times 18.5 = 3.885 \text{ hours per } 1,000.$$

If more than one person is working at a station, the hours per 1,000 is multiplied by the number of people:

$$2 \text{ people} = 7.770 \text{ hours (twice as much time)}$$
$$3 \text{ people} = 11.655 \text{ hours (three times as much time)}.$$

Hours per 1,000 equals the 100% station average time times 18.5 hours/1,000 times the number of operators at that station for every station. Every work station will be 3.885 hours per 1,000 or multiples of that number. This is what line balance means—

everyone works at the same pace. The logic for a work station with three people taking three times as many hours per 1,000 should be obvious. Another piece of logic is that everyone on an assembly line must work at the same rate. The person with the least work to do still cannot do one more than comes to the operator and cannot do one more than the following operator can do.

⑮ Pieces/hour: Pieces per hour is $1/x$ of hours/1,000 times 1,000 (or divide the hours per 1,000 into 1,000). Notice in our example, Figure 15-1, that all the stations produce 257 pieces. Station 05 has two operators, each producing 129 pieces per hour for 258 pieces per hour total.

⑯ Total hours/1,000: Total hours per 1,000 is the number of hours adding all the operations together. The hours per 1,000 for one operator times the total number of operators on the line also equals the total hours per 1,000. The total of column ⑬ is the total operators.

⑰ is an average hourly wage rate. This would come from the payroll department, but let's say $7.50/hour is the average hourly wage rate.

⑱ is the labor cost per 1,000 units. In our example, 46.62 hours times $7.50/hour = $349.65 per 1,000 units of $.35 each labor cost. The lower the cost, the better line balance.

⑲ is the total cycle time. The total cycle time tells us the exact work content of the whole assembly, and if treated like any other time standard can show us what a perfect line balance would be.

In our example, 1.980 minutes × 18.5 hours/1,000 equals 36.63 hours per 1,000. Our line balance came out to 46.62—10 hours more. This 10 hours is potential cost reduction, and what cannot be removed by cost reduction is called the cost of line balance.

IMPROVING ASSEMBLY LINE BALANCING

The end of an initial line balance will lead to improvements. To improve line balance, we look at

1. Reducing the 100% station. This can be done by
   a. adding an operator
   b. cost reduction.
2. Combining the 100% station with an operation in front of or behind, but sequence of operations must be maintained.
3. Combining other operations together to eliminate one of them.

FIGURE 15-3 Assembly line balance example

⑦	⑧	⑨	⑩	⑪	⑫	⑬	⑭	⑮
						Ave.		
Operation No.	Operation Description	R	Cycle Time	No. Stations	% Load	Cycle Time	Hours/ 1,000	Pieces/ Hour
5	Assemble	.216	.357	2	.179	100%	6.623	151
10	Assemble	.216	.441	3	.147	82%	9.935	101
15	Cement	.216	.210	2	.105	59%	6.623	151
20	Rivet	.216	.344	2	.172	96%	6.623	151
25	Form carton	.216	.166	1	.166	93%	3.312	302
30	Label	.216	.126	1	.126	70%	3.312	302
35	Packout	.216	.336	2	.168	94%	6.623	151
Total			1.980	13			43.051	
							⑯	

Let's do our example over again (Figure 15-3) with the addition of one more operator at station 15, the 100% station.

The original line balance was 46.62 hours. Subtracting the new line balance of 43.05 shows a savings of 3.57 hours, and at the rate of $7.50/hour, the savings are $28.78 per 1,000, or about $50/day or $12,500 per year. This is still not a good balance. Can you do better?

SPEED OF THE CONVEYER LINE

On the first day of production, someone is going to ask you how fast to run the conveyer belt. Conveyer speed is a combination of product size and "R" value.

Conveyer speed is measured in feet per minute, and when an R value is determined, a finished part must roll off the assembly line at that speed. R values are normal time, and the line speed is calculated or based on normal time. The R value example of .250 minutes is a good one: .250 minutes is four parts per minute, and that is the number of finished units to cross the finish line every minute. Since a conveyer moves at the same rate from the first operator to the last operator, four parts per minute will be the speed at every work station. The only other piece of information needed is the length of a finished product. For example, a swing set is 10 feet long. A swing set coming down the line is end to end and takes up 10 feet of belt. So we need four swing sets per minute at 10 feet each, or 40 feet per minute belt speed. A 2-foot part at four per minute equals 8 feet/minute. Remembering our time standard for walking (264 feet/minute) from Rating in chapter 9 helps us develop the proper perspective about these belt speeds.

No allowances are in belt speed, because the line will be off during breaks and delays. So if allowances were added to belt speed, we would never achieve 100% performance.

At the rate of 4 pieces/minute = 240 possible per hour
.250 minute × 18.5 = 4.625 hours/1,000 = 216 pieces per hour

Our standard is 216 pieces per hour, but our belt speed is for 240 per hour. The extra parts will be lost when the belt is off. Two hundred and sixteen parts per hour × 8 hours per sift = 1,728, but we only wanted 1,200 units per shift. No plant operates at 100% performance. If this is a new product, the average performance during the first year will be closer to 70%.

70% of 1,728 units = 1,209 units per shift. Is that close enough?

Examples

1. Initial try (Figure 15-4)
2. Improved solution (Figure 15-5)
3. Plant layout (Figure 15-6)

FIGURE 15-4 Assembly line balance: Initial try

FRED MEYERS & ASSOCIATES — ASSEMBLY LINE BALANCING

PRODUCT NO.: 1670
DATE: 10-10-xx
BY I.E.: F. Meyers

PRODUCT DESCRIPTION,
New plastic charger
NUMBER UNITS REQUIRED PER SHIFT 1200

"R" VALUE CALCULATIONS

EXISTING PRODUCT = $\frac{365\ MINUTES}{UNITS\ REQ'D/SHIFT}$ = 'R'

NEW PRODUCT = $\frac{300\ MINUTES}{UNITS\ REQ'D/SHIFT}$ = 'R'

NO.	OPERATION/DESCRIPTION	'R' VALUE	CYCLE TIME	# STATIONS	AVG. CYCLE TIME	% LOAD	HRS./1000 LINE BALANCE	PCS./HR. LINE BALANCE
1	Place bottom housing on line and lub	.250	.200	1	200	80	4.17	240
2	Assemble parts 3, 4, & 5	.250	.250	1	250	100	4.17	240
3	Assemble parts 6 & 7 together and Place sub-assembly in housing	.250	.305	2	153	61	8.34	120
4	Drive 6 bolts holding sub-assembly in bottom housing	.250	.600	3	200	80	1250	80
5	Get vent cover & cement in place in bottom housing	.250	.198	1	.198	79	4.17	240
6	Get top housing, apply cement & assemble to bottom housing	.250	.290	2	.145	58	8.34	120
7	Place in carton & in master carton (6 per) and aside to pallet	.250	.625	3	.208	83	12.50	80
							54.19	
	EXAMPLE AND ASSIGNMENT:						x $10.00/hr.	
	Improve this line balance – Reduce cost.						$541.90/1000	
	Your goal is to make this product as cheaply as possible.						$ .54 each	

FRED MEYERS & ASSOCIATES — ASSEMBLY LINE BALANCING

PRODUCT NO.: 1670	PRODUCT DESCRIPTION,	
DATE: 10-11-XX	NEW PLASTIC CHARGER	
BY I.E.: MEYERS	NUMBER UNITS REQUIRED PER SHIFT 1200	

"R" VALUE CALCULATIONS

EXISTING PRODUCT= $\frac{365 \text{ MINUTES}}{\text{UNITS REQ'D/SHIFT}}$ ="R"

NEW PRODUCT= $\frac{300 \text{ MINUTES}}{\text{UNITS REQ'D/SHIFT}}$ ="R"

NO.	OPERATION/DESCRIPTION	"R" VALUE	CYCLE TIME	# STATIONS	AVG. CYCLE TIME	% LOAD	HRS./1000 LINE BALANCE	PCS./HR. LINE BALANCE
1 & 2	Place Bottom Housing on line, Lubricate & assemble parts # 3, 4 & 5.		.450	2	.225	98	.00763	131
3 & 4	Assemble Parts # 6 & 7 Together, Place in Housing & Drive 6 Bolts Holding sub to Housing.		.905	4	.226	99	.01526	65
5	Get Vent Cover & Cement in Place in Bottom Housing.		.198	1	.198	86	.00382	262
6 & 7	Get Top Housing, apply cement & assemble to Bottom Housing, Pack Out.		.915	4	.229	100	.01526	65
							.04197	

FIGURE 15-5 Assembly line balance: Improved solution

CONCLUSION

Line balancing is an important tool for many aspects of industrial technology, and one of the most important uses is the assembly line layout. The back of the assembly line balancing form is designed for an assembly line layout sketch. Look at the example in Figures 15-4, 15-5, and 15-6.

Packout work is considered to be the same as assembly work as far as assembly line balancing is concerned. Many other jobs may be performed on or near the assembly line but are considered subassemblies and are not directly balanced to the line (because subassemblies can be stockpiled). Their time standards stand on their own merit.

QUESTIONS

1. What are the purposes of assembly line balance?
2. What information is required before an assembly line balance can be tried?

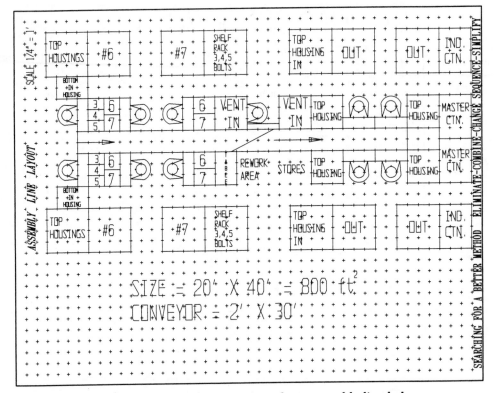

FIGURE 15-6 Assembly line layout resulting from assembly line balance in Figure 15-5.

3. What is an R value?

4. What is the average cycle time of ⑫?

5. What does % load mean in ⑬?

6. What does 18.5 mean?

7. Why is only the 100% station standard used?

8. What is the significance of total hours per 1,000 in ⑯?

9. What is the significance of total cycle time in ⑲?

10. Improve the line balance that appears in Figure 15-3.

11. When should subassemblies be placed on the assembly and packout line?

12. What should the speed of the conveyer be with the following?

R	Length	Line Speed
.162	96″	
.440	24″	
1.100	6″	

How would incentives change the above?

13. Improve the line balance of the example in Figure 15-4. How much can you save?

14. Review the layout in Figure 15-6.

15. How does the assembly line balance relate to the line layout?

CHAPTER 16

Indirect Labor and Motion and Time Study

The number of people and the cost of indirect labor are determined for budgeting using ratios of indirect labor categories to direct labor based on last year's actual head count. (Review Figure 16-1 for examples of indirect labor.) If past practices are good enough, then this plan is good. But should we be satisfied with present cost, methods, procedures, systems, or manpower? As direct labor is worth the effort of motion and time study, so is indirect labor.

The techniques of motion and time study used for direct labor are also used for indirect labor. In this chapter, we discuss indirect labor in detail and provide examples of how categories of indirect labor were affected by motion and time study.

The example of indirect ratios in Figure 16-1 shows what type of work is considered indirect, and the ratios indicate national industrial averages. Twenty-one percent of direct labor (210/1,000) or 17.4% of the total people (210/1,210) are indirect. Many of these indirects earn more than direct labor, so cost percentages will be higher.

When a new budget is prepared for next year, these ratios will be used. If business is good and more people are hired for direct labor, the indirect labor also fluctuates, but not as fast. If business conditions call for a layoff, the ratios must be maintained—meaning a reduction in each category. Ratios are very important, and I would not eliminate them. However, the material handling category in Figure 16-1 is quite large, and the potential savings can be spectacular. Another problem with ratios occurs when incentive systems are installed. Now the direct labor people are working 42% faster (using 42% more material per unit of time) with the same crew size. Do we add more people, or do we put material handlers on incentive? Are we confident that material handling

FIGURE 16-1 Factory labor analysis: Last year

Category	No. People	% of Direct Labor
1. Material handling/control	60	6%
2. Quality control	30	3%
3. Manufacturing, plant, and industrial engineering	14	1.4%
4. Supervision/management	45	4.5%
5. Maintenance/tooling	17	1.7%
6. Warehouse/shipping	18	1.8%
7. Receiving/stores	7	.7%
8. Factory clerical	10	1.0%
9. Miscellaneous	9	.9%
10. Direct labor	1,000	
11. Total	1,210	21.0%

equipment can do 42% more work? These questions are most difficult to answer if management does not know what the indirect people can do.

MATERIAL HANDLING

Methods work (motion study) can improve material handling more than anything else. A few points to consider are as follows:

1. Can we eliminate any moves?
2. Can we automate any moves?
3. Can we combine operations to eliminate the moves between stations?
4. Can we move more parts at one time?
5. Can we reduce transporting empty by planning backhauling?
6. Can we manufacture a part in-line using gravity feeds?
7. Can we manufacture the part next to the assembly line packer or assembler?
8. Can we move machines closer together?
9. Are we using the best material handling device?
10. Are we using stand-up reach trucks or narrow aisle trucks instead of sit-down fork trucks?
11. Can we use moving storage—overhead, tow, gravity chutes, skate wheel, or roller conveyer?

After the technologist has reduced material handling to the minimum economically feasible, time standards can be set. Each move must be listed (each part and the container count) and the distance estimated (averages are usable). A foot-per-minute travel time can be determined (remember, a person walks 264 feet per minute) for each piece of equipment, and a time for maneuvering (set down, pick up) can be determined. Any miscellaneous job can be time-studied or an allowance can be built into the estimated standard.

The objective of material handling standards are not performance reports but only to see if the person is properly loaded with work.

Standard data is the most economical time standard, and once the data is built, application is very fast. The job standard only needs to be reviewed when the work load changes.

QUALITY CONTROL

Quality control varies greatly from one company to another, and the need for people varies considerably. The highly labor-intense operations of quality control are the main subject of this section.

1. Line Inspectors: Many assembly lines have inspection permanently assigned to the end of the line. They have work stations, tools, and gauges just like any other operator. The motion and time study technique of PTSS allows the technologist to develop the most efficient method and to set a time standard for that method just like any other job. On assembly lines, the inspector is balanced just like any other operator.

2. Department Inspectors: The purpose here is to know if proper manning is allowed for. A department inspector will roam the area randomly checking material and operations. Department inspectors may be required to approve set-ups before the operator starts production, and they may be required to inspect a part every hour. The requirements must be determined and elemental inspection times applied to each.

3. Inspection Department: Material is moved to the inspection department for 100% inspection. This is the least productive method of inspection, but special tooling and gauges may dictate this type of inspection. Motion and time study of this type of inspection is exactly like any other mass-production job.

MANUFACTURING, PLANT, AND INDUSTRIAL ENGINEERING

These highly professional people need goals, too. The expert opinion standards and the backlog control system can be effectively used in these areas. Methods improvements are also valuable in these areas. Standard data is the most efficient method of setting standards, and the number of standards set per manhour is a good indication of efficiency. In consulting, I'll bid a job using one half hour per time standard.

Plant engineering can be made efficient by use of CAD systems showing overlays on the plant layout for air, water, electric, heat, air conditioning, lighting, etc. Inventory of equipment and tooling can include expenses of repair and justify replacement.

Manufacturing engineering can use computer aided design (CAD), computer aided manufacturing (CAM) for tool and fixture design. New software is available for sourcing tools, machine, repair parts, and components needed in the design of new facilities.

SUPERVISION

Supervision can be assisted by good methods work, locating supervisors close to their people and the services needed, and designing good systems and procedures for controlling their needs. A supervisor ratio of 20 to 25 people per supervisor in manufacturing is good. This ratio drops to 10:1 in offices, engineering, and upper management. Supervision is probably one area where time standards would be inappropriate. Ratios are useful.

MAINTENANCE AND TOOLING

Maintenance and tooling cannot be talked about in total. Maintenance and tooling is made up of routine maintenance, ordinary maintenance, and emergency maintenance. Each category must be handled separately.

1. Routine maintenance is often time called scheduled maintenance, because the maintenance is required on a specific interval and needs to be scheduled. Oiling machinery, changing bearings, bulbs, oil, coolant, and inspecting equipment is a small list of routine maintenance chores that can be time-studied and scheduled. Using the computer, a list of jobs could be printed out each morning for each routine maintenance person, giving them 8 hours of work. If the routine maintenance person finds something else wrong, he or she writes up a work order for ordinary maintenance or calls for emergency help at once. However, the routine maintenance person must move on to the next job.

2. Ordinary maintenance (planned maintenance) is the kind that can be planned for, scheduled, and completed in a modest amount of time. Building new things, modifications to existing facilities, or major repairs fall into this category. The expert opinion time standard system and backlog control system are used on this category of maintenance.

3. Emergency maintenance needs to be done immediately. Any qualified craftsman working on ordinary maintenance jobs can be called away from that job to get the plant running again. Emergency maintenance is the most expensive maintenance, because people are waiting and there is no time to plan. For this reason, emergency maintenance must be discouraged and replaced. Good routine maintenance and inspec-

tions can predict breakdowns and be corrected as ordinary maintenance projects. A performance report for maintenance would surely include how much emergency maintenance was performed with the goal of elimination. The newsprint business has nearly eliminated breakdowns by predicting and repairing problems before the breakdown. The following are a few methods for improvement in maintenance:

1. Vibration analysis
2. Heat recordings
3. Machine records
   a. Type of repair
   b. Oil consumption
4. Automatic lube systems
5. Instrumentation for flow, temp, amp, etc.
6. Replacement part inventory with reorder points and reorder quantity
7. Plug-out/plug-in replacement units
8. Back-up equipment.

WAREHOUSING AND SHIPPING

Warehousing is the safekeeping and issuing of the company's finished product. Great work has been accomplished in warehousing methods. A few examples of warehousing philosophies are as follows:

1. Locate a small amount of everything is a small area to reduce the travel time to pick orders. Locate back-up stock close to picking area.
2. Have a fixed location for every finished product to aid pickers learning the locations of parts.
3. Place the highest-volume (most popular) products in the most convenient locations.
4. Design the location of all products to be immediately available without having to move other products.
5. Place the highest-volume products closest to the shipping door.
6. Pick orders and pack at the same time.
7. Design all material handling equipment to work together—for example, picking carts the same height as packout conveyers for rolling material off the cart and onto a roller conveyor.

A major brand tool company has six product lines (different name brands) and over 5,000 different stock-keeping units (SKUs). The tools were stored on seven high shelf racks, 3 feet wide, in part-number order. Thirty-six back-to-back rows of shelves,

30 feet deep, were needed. A person picking an order of tools needed to pass in front of every shelf, a trip of over 2,100 feet. The new layout placed the tools in order of popularity and classified each tool as an A, B, C, or D item. A items accounted for 80% of the sales dollars but only 20% of the part numbers. These were very popular tools that moved fast. These items were placed close to shipping. On the other end of the scale, D items were slow-moving tools and accounted for 5% of sales but 40% of part numbers. These tools were located at the storage area farthest away from shipping, because the picker needed only go that far for 5% of the orders. B and C items were in between. The travel distance was reduced to 350 feet, a savings of 1,740 feet/order. This savings is astronomical, but to put a dollar figure on it, time standards are needed. In most of warehousing, three things cause time to vary:

1. The number of orders picked;
2. The number of lines per order; and
3. The number of individual items per line.

In the tool warehouse, the warehouse person received an order at the supervisor's desk and gave it to an inspector. This work took two minutes. Between each line, the operator had to record the number of items picked and walked to the next location. Average distances were used to develop a standard of .3 minutes per line. Once at the location, an operator could pick a unit every .2 minutes (a unit may have been one socket, but they were packed in boxes of six). A time card would look like the example in Figure 16-2.

Consider another example. An oil company warehouse was set up according to the A, B, C inventory classification system. The pickers drove a fork truck. Cases of oil were the units of sale. (See Figure 16-3).

FIGURE 16-2 Daily performance report: Picker #12

	Order No.	No. Lines	No. Pieces
1.	123	25	100
2.	149	55	300
3.	175	28	250
4.	201	35	150
5.	222	15	500
6.	251	45	400
7.	300	10	600
Total	7	213	2,300
	$\times$	$\times$	$\times$
Standard	2.0 minutes	.3 minutes	.2 minutes per piece

$$\text{earned minutes} = 14 + 63.9 + 460 = 538$$

$$\frac{538 \text{ earned minutes}}{480 \text{ actual minutes}} = 112\%$$

FIGURE 16-3 Daily performance report: Oil warehouse

	Order No.	No. Lines	No. Cases	Earned Minutes
1.	167	30	250	74
2.	250	1	900	95.5
3.	950	20	400	74
4.	1295	15	300	56.5
Total	4 orders times	66 lines times	1,850 cases time	
Standard	4 minutes/order equals	1.5 minutes/line equals	.1 minute/case equals	

Earned minutes   16 minutes + 99 minutes + 185 minutes = total earned minutes, of 300

800 cases makes one truckload (sixteen pallets/trailer)

$$\text{Efficiency} = \frac{300 \text{ earned minutes}}{480 \text{ actual minutes}} = 62\tfrac{1}{2}\% \text{ performance}$$

Shipping, on the other hand, usually includes closing containers, weighing containers, addressing containers, writing the case number weight and carrier's name on the shipper, making out bill of lading, and loading trucks. Individual time standards could be calculated for each of these functions, but a much easier way is to keep a count of the weight being shipped every day and divide this by the total manhours used by everyone in the shipping department.

$$\text{Performance} = \frac{480,000 \text{ lb.}}{48 \text{ manhours}} = 10,000 \text{ lb./manhour}$$

If we calculate this daily, a meaningful trend line should soon develop. This is a standard on which one could base a 50/50 incentive plan. A year's worth of history would provide a great base standard.

Drivers

Some companies use their own drivers. The oil company warehouse example given earlier had twenty drivers. A driver performance report was based on 50 mph on the highway, 25 mph in the city, 20 minutes per stop, plus 2 minutes per pallet. The drivers were loaded with a known amount of work and were expected back at a specific time for their next load or inside work.

A lumber company in Portland, Oregon had thirty-three trucks and drivers delivering lumber. All thirty-three were loaded by yard crews overnight, and at 7:00 A.M. all trucks left the yard to make their first run. At 10:00 A.M., trucks were lined up waiting for their next runs to be loaded. The drivers were idle during this loading time. A methods improvement was to eliminate six drivers and use their trucks as back-ups to the others. When a driver returned from a delivery, he or she left that truck and picked up the next loaded truck in line—no waiting. Time standards were developed for each run by the dispatcher, and an estimated time of arrival (ETA) was placed next to that driver's name on the dispatch board. Drivers had no problem if they were late once in a while, but consistent lateness earned them some coaching. This may sound a little hardnosed, but this company saved over $300,000 per year with this system.

RECEIVING AND STORES DEPARTMENT

The receiving department's responsibilities are to unload carriers, create a receiving report, and check in all material as to quantity, quality (visual only), and correctness of part numbers. Receiving is a smaller department than shipping, but it can develop a pounds-per-manhour standard or a trucks-received-per-manhour standard. Sometimes companies combine receiving and shipping together.

The trucking industry delivers products coming into the area in the morning and picks up finished products from shipping in the afternoon. This is called less-than-truck-load (LTL) shipping or common carriers shipping. All the less-than-truck-load freight for your company is delivered to a central trucking company station in your town and is collected until the next morning. The local common carrier will load your material and several other companies' materials onto a trailer and make its deliveries. This high-morning work load for receiving is combined with a high-afternoon work load with shipping, making one easily manner combined department. The combined department will use pounds shipped per manhour as a standard because everything shipped must have been received, and in the long run, shipping and receiving must be equal.

Common equipment is used in both shipping and receiving; therefore, fewer pieces of expensive material handling equipment may be needed if shipping and receiving are combined. Fork trucks, scales, and dock doors are a few examples of this expensive equipment.

Employee qualifications are similar for both receiving and shipping: attention to detail, equipment operations, and responsibility for company assets.

Stores' responsibilities include locating, safekeeping and delivering production materials to the first production operation. A stores clerk must put stock away in such a way that it can be readily available when needed, and must retrieve that stock when ordered by production. Storekeepers must keep accurate records of locations and inventory on hand.

All movement of material must be authorized to insure inventory records are correct. Unauthorized movement of material could cause a plant shut down because the material needed is not there even though the inventory list said it was.

Methods improvement ideas for storerooms include the following:

1. Random locations: Put the pallet or box into the first open space the driver comes to and record that location on a locater ticket. This reduces driving time and reduces the storage space requirements of the store room by one half.

2. Create addresses for every location in the warehouse. Each row could be a number, each pallet location back from the main aisle could have a letter (A, B, C), and each shelf would have a number (floor 1, 2, 3 4). An 11-F-4 location would be the eleventh row of racks, the sixth pallet location down that row, and on the top tier. F would always be the sixth location back, and four would always be the top.

3. Locate all slow-moving and obsolete inventory in the back of the storeroom.

4. Use narrow aisle vehicles to save space.

Time standards for storekeepers are based on the number of pallets put away and the number of pallets retrieved. With a little planning, one can be put away and one can be retrieved on every round trip, making for great efficiency. A locater ticket must be made out for every item put away, and a locater ticket must be pulled for every item withdrawn from the storeroom. These locater tickets can be collected and multiplied by a standard hour per move to develop an earned hours figure.

$$\text{Efficiency} = \frac{\text{earned hours}}{\text{actual hours}}$$

FACTORY/CLERICAL

Many factory clerical people are staff assistants to busy factory managers and are an extension of the manager. These positions are justified by saving the manager time for more important duties. No attempt will be made to measure their performance, only their supervisor's performance.

Other factory clerical people may be required to keep records of activities, inventory, equipment, cost, etc. These activities and transaction times can be determined. Otherwise, how do we know if the person has enough work—or too much?

Computer input clerks are often monitored by their computers, and key strokes per hour are printed out each day for each person. This is very useful data, but the different job must have different expectations (standards).

OTHER INDIRECT AREAS

1. Utilities companies
2. Mining companies
3. Hospitals
4. Retail stores
5. Insurance companies
6. Airlines
7. Hotels

8. Restaurants

9. Anything not here will be a good future opportunity.

QUESTIONS

1. What is direct and indirect labor?

2. How are numbers of indirect people calculated for budgets?

3. List eight areas of indirect labor.

4. List two methods improvement ideas for each category of indirect labor.

5. When would you use the stopwatch, PTSS, or expert opinion time standards?

CHAPTER 17

Time Management Techniques

INTRODUCTION

College students in their junior year have had to become good time managers to get that far. Every student has experienced the pleasure of accomplishing superhuman amounts of work in limited amounts of time when he or she had to. What if we could produce at this high level 52 weeks per year? You may consider this funny now, but successful people will do exactly that. The time management techniques outlined in this chapter will help you produce many times your normal results.

Two truisms are important to us:

1. If you want something done on time, give it to a busy person.
2. There is always time to do what you truly want to do.

The second truism tells everyone your attitude about those things you didn't do, so be careful.

The industrial technologist's time is valuable, and you should know the dollar amount of your time. This will keep you from spending your time on work not worth the cost. Knowing what your time is worth and doing only what is worthwhile is an attitude that will make you more valuable. Time management techniques are ways of getting more out of life. They are tools to help you determine what's important and to keep you from wasting time. There are eight basic time management techniques:

1. Creativity
2. Selectivity
3. Delegation

4. Do it now
5. Set time standards for yourself
6. Eliminate the unnecessary
7. Respect other people's time
8. Do not rationalize.

CREATIVITY

Creativity is finding new ways to do the work. The industrial technologist is great at finding better methods for others but should also think about his or her own work. Some ideas include the following:

1. Word processors for periodic reports
2. Spreadsheet programs for standards and cost reduction
3. CAD for layout and work station design
4. Standard data for time standards
5. Forms and formulas for costing
6. Computer time study for establishing standard data
7. On-line performance reporting
8. Downtime monitoring.

Whatever the job, we can eliminate, combine, change sequence, or simplify that job to save time and money. The more routine the job, the more we can save.

SELECTIVITY

Selectivity is important in every area of human endeavor. Every engineer and manager must make decisions on what is the most important to them at this moment. Job problems, family problems, and social problems may all occur at the same time. Which one should you work on first? There is no one rule; each situation is different, and the choice is yours alone. The problem comes when selections are made without all the information needed to set priorities.

The industrial technologist must know what is important and give the most important tasks top priority. Knowing what is important requires an understanding of what the boss wants and the other work demands.

The boss's requests take top priority. At times you may not agree with this, so talk to the boss. The boss may not know what you are doing and that it is important, so discuss priorities. A good relationship with a supervisor starts with good communications, and telling your boss what you think will make life easier for both of you. I would suggest you tell the boss, "OK, I'll do it right away, but do you know I've been asked to do so and so? Should I drop that for now and do what you are asking?"

The routine and special jobs that make up your backlog of work should be listed and prioritized. When additional projects are added to your backlog, be sure you understand the priority. Once all jobs have been listed, then do the most important job first.

Take notes at meetings, and add requested or promised work to your backlog list. Do what you promise and do it on time.

Learn to say no. People will ask you to do things which you should not do because of various reasons: Maybe someone else could do it better, maybe it is social (coffee break) and you don't have time, maybe it is some club they want you to join, or maybe it is some community service project. The more successful you become, the more demands will be made on your time. There will always be only 24 hours per day, so be selective and learn to say no diplomatically.

Selecting what is important and concentrating on that problem will make you a valued employee. Keeping a proper balance between work, family, and social demands will make you a successful person. Good time management techniques application will allow you to get more out of life.

DELEGATION

Delegation is assigning other people some of your work. As your work load grows, your employees can lend assistance. As your ability to solve problems increases, more problems and responsibility will be added. Management rewards results with more sources, a secretary, an assistant, a department, etc. With the addition of the first part-time help, you must decide what you are going to give away in the form of work.

Routine tasks are the first jobs to delegate. Daily, weekly, or monthly reports, collecting routine information, or providing routine services can be delegated.

Priorities are set for all projects. ABC priority designations could mean the A priorities are the ones you do, and you will delegate B and C priorities because they are not as important.

Screen calls and mail. Much time is wasted on meaningless phone calls and junk mail, so delegate answering the phone and opening the mail to someone else. Develop in this person an ability to prioritize.

Enjoy what you do. Do what you do best and recruit people to complement and augment your skills, so as not to duplicate yourself. This is called team building.

Whenever you delegate anything, you give up the work but not the responsibility. Keep informed on what has been delegated through briefings.

DO IT NOW

W. Clement Stone, a Chicago self-made billionaire, is known for saying, "Do it now!" He says it at the top of his voice. An application of this statement is, "Don't put that project down until it is finished.'' An example is opening a report and reading it now, instead of putting it in a basket to read it later. Read it now, and throw it away or file permanently. Putting things off (procrastination) is a disease. Like the IRS recommends—do your taxes early and stop the worry; it's not so bad. Many unpleasant jobs cause us more headaches and heartaches when put off. Do it now, and stop worrying.

Many of us have put something off so long that it is embarrassing when we finally do it. Writing, returning something borrowed, returning a phone call, turning in a report, etc. are familiar examples. Doing it now is a habit that will save time because we handle, think, and worry about it only for a short time.

SET TIME STANDARDS FOR YOURSELF

Time standards are goals, and each new project should be estimated and the time recorded. These time standards are expert opinion standards and do not need to be as accurate as direct labor standards. Time standards will limit the amount of time given to any project. Sometimes we study a problem too long, and a time standard tells us to move on.

The backlog control system discussed in chapter 12 can be very useful to the industrial technologist. The backlog hours tells management how much work needs to be accomplished. If this time frame is too long, management will provide help. The backlog control graph is the best way to communicate to management just how much work we are doing. The backlog control system will save much more time than it takes to complete the forms. Systematic approaches to management are always better than seat-of-the-pants approaches.

Time standard estimates are another way of ensuring communications between you and the boss. The boss may have been thinking about a 15-minute project, and your estimate of 3 days indicates a misunderstanding. Cost and relationships can be saved by developing and communicating time standards.

The most important employee you will ever need to control is you. Set time standards for your work.

ELIMINATE THE UNNECESSARY

Knowing what is unnecessary is as important as knowing what is most important. We want to eliminate insignificant tasks and nonproductive activities. Insignificant tasks are jobs that would be more economically done by other employees at lower pay rates. Preoccupation, remembering errors, and planning too much are examples of nonproductive activities, and they waste your time.

Preoccupation can take many forms: personal problems, marital problems, financial problems, people problems at work, and past errors are all examples of problems that even successful employees can use themselves in. Preoccupation must be controlled for your own piece of mind. The best rule is, "Do what you can now and move on to the next project."

Remembering past unpleasant problems is also a waste of time and energy, and it harms your attitude in the present situation. Put mistakes behind you as soon as possible. Learn from the mistakes, talk about them with your supervisor, but do not waste one moment worrying about how you didn't do it right. No one is perfect, everyone has made mistakes, and you will make mistakes again. We must try not to make the same mistake twice. Talking with the boss is important. Tell the boss what you did and how bad you feel, and then ask if you can go on from here.Taking responsibility for your actions is a sign of maturity, and every supervisor appreciates this.

Thinking ahead too much can also be nonproductive. Planning is good and it is needed, but planning for every consequence of every action will only slow you down. People that plan every detail make very few mistakes, but they also do very little work. The challenge for every technologist is to know how much planning is enough. At some point, planning must stop and implementation must start.

RESPECT OTHER PEOPLE'S TIME

If you were deep in concentration and someone stopped to talk about last night's game, your concentration has been disrupted. Much time has been lost getting back to where you were before the interruption. Socialization is important to a well-balanced life, but keep it to social times (lunch, breaks, happy hour). Before and after regular hours are good times to be social with the boss.

When you visit other people's offices, check with them or their secretary for a meeting time. Set up some signals like open door, office hours, or schedules with secretaries for meeting times.

Don't expect the boss to do your work. Don't bounce everything off the boss. Get good instructions up front and turn in completed reports. If the boss doesn't like something, he or she will let you know. You can let the boss know what you are doing by well-planned project status reports. Keep the boss informed, but respect his or her time.

DO NOT RATIONALIZE

Rationalizing is coming up with a good excuse for doing something wrong. Excuses are antiproductive and a waste of everyone's time. Doing something unnecessary because you do it so well is rationalizing. Putting something off because someone did not do what they were supposed to is rationalizing. Not doing something the boss thought was important because you did not have time is rationalizing. There is no good excuse other than your inability.

Time management techniques will allow you do more in the same amount of time. You can get $1\frac{1}{2}$ years of experience every year because you manage your time. You will go farther and faster than your fellow employees because you have more time.

QUESTIONS

1. What are the eight basic techniques of time management?
2. Give two personal examples of each.
3. Is there *always* time to do what you truly want to do? Why?

CHAPTER 18

Attitudes and Goals of Industrial Technologists

An industrial technologist solves problems and reduces costs. The industrial technologist works with everyone in the organization and must possess human relation skills. No graduate of a time management course has all the information needed to deal successfully with the wide variety of problems in industry today. To help the new industrial technologist cope with these problems, and to keep him or her from reinventing the solutions to common problems, we discuss attitudes and goals in this chapter. Try some of these attitudes on prospective employers. You will be grateful to know that this is what they are looking for.

To achieve anything, we needs goals, and to achieve goals, we need the right attitudes. Control your attitudes, and you will achieve every goal.

ATTITUDES OF AN INDUSTRIAL TECHNOLOGIST

I Can Reduce the Cost of Any Job

1. I can eliminate the unnecessary.
2. I can combine jobs together and eliminate the movement, storage, and some handling time.
3. I can change sequence of work elements around making a more efficient operation.
4. I can always simplify the job by moving things closer and downgrade work elements to produce more work in less time with less effort; there is always a better way to do any job.

Cost is Number One

Decisions are made to reduce total cost, and the deciding factor is a combination of savings (return) and expense (investment). The return on investment must be sufficient to motivate management to take a chance on your idea. A 100% return on investment will pay itself back the first year, and 100% is the most common objective for methods work.

The savings (return) is from labor standards created from before and after motion studies. The investment is the total cost of implementing the methods improvement. The best methods improvements are those that reduce cost with no investment.

Growth Gives Meaning and Purpose to Life

Never stop learning. When starting a new job, the product, people, systems, and procedures are all new. Learn as much as you can about the company.

Industry is getting more complicated every day. An industrial technologist must keep up with the growth of his or her profession and industry. Become a member of a professional society, attend its annual conferences, and read its monthly publications. Obsolescence of an industrial technologist takes years, but it starts the day you graduate from a time management course.

Problems are Opportunities

Problem solving is the job, and a proper attitude toward problems helps keep the technologist going. Problem solving is also how the technologist keeps growing professionally. The more problems a technologist can solve, the more management depends on him or her to solve more and bigger problems. Big problems are the greatest opportunities to show top management what you can do. There are people who hate and avoid problems. There are routine tasks for these people, but not top engineering and management positions.

I Work at 110%; I Work at 125%

Productivity improvement and work ethic are two attitudes the industrial technologist must sell at all times. "Do as I do and as I say" is an easy sell and demonstrates what you expect of yourself. Your work pace is a habit and will not go unnoticed by upper management as well as shop employees.

Anything Done the Same Way for More than a Year is Obsolete

Selling change is a tough job and must be done continually. Every time you get a chance to promote change, do so. Companies do not live in a vacuum. A company is in competition with other companies around the world, and if it allows itself to be satisfied with the present methods, it will lose the competitive fight for jobs. The company must become better every day or it starts to die.

The Person Doing the Job Knows More About It than I

The opposite view is not only wrong, it is stupid. The person on the job has worked on it for a long time, probably 8 hours per day, and is a wealth of information. The industrial technologist that recruits the operator's help will be many times more effective than the technologist working on his or her own. Asking the operator's opinion also shows respect, and we all want to be respected.

The Best Ideas are Developed in Groups

One person cannot know as much as two, and if people will talk with each other, there is no limit to what can be done. American management has grown up with the lone ranger complex. Today's world is too complicated for the loner.

Education is a Part of My Job

Motion consciousness and cost consciousness are two areas of expertise we can share with others, and if we teach people what we are doing, they will accept our suggestions easier and faster. Also, we will motivate others to do as we do.

I Will be Open and Honest with Everyone

Having ulterior motives and playing games are dangerous to good relationships. You can disagree with people without being disagreeable. You can say "I disagree with that thought" while touching the person's arm (meaning "I like you").

Tell people you are the most honest person they know, and of course, act that way.

Tell people you would do nothing in the world to hurt them, and mean it. Help them whenever you can.

Quiet, inward people will have a difficult time being an industrial technologist.

All My Standards and Goals are Attainable

Be confident in your standards. If you don't believe in them, nobody will. Later in this chapter, we discuss the only way to set time standards—the right way, which is fair both to the company and to the employee.

I Will be Patient and Understanding Toward People

The motion and time study job has a history of antagonism with production employees. Their attitude was developed years ago by others in the profession, but you understand that this is not personal and you will do better working with these people. You are the most honest person they know, and you tell them so.

If They Don't Like Me, They Don't Know Me

We cannot get a group of people to know and like us; we can only work on one at a time. If you have the proper attitude toward people, and if we do our job professionally (honestly), people will like us—one at a time. People must know you as a friendly, happy person.

If People Don't Understand My Ideas, I'm Not Communicating

I take responsibility for communicating my ideas by telling and listening to feedback. Many techniques of motion and time study perform the same function and communicate the same information. By using two or three techniques we may be able to communicate better with more people. If I take responsibility for the communications, I'll try harder. Eighty percent of engineering and management is selling ideas, and this is the most overlooked part of industrial technology education. The successful people are the ones who can sell their ideas.

No One Wants to Know How Good You Are, But They Want to Hear How Good They Are

A conversation with either employees or supervisors should be directed toward them and not you. You will be considered wise and all knowing when you listen to and applaud their accomplishments. This is especially true with bosses.

Successful People Do what Other People Dislike to Do

1. They work hard. They know that there is no easy way. They collect all the data before making decisions, ask for help, and take advice.

2. They work long. You can compete successfully with people who work fewer hours than you, and over the years of working longer you build up a large lead on the competition for a big new job. The boss has learned this and is usually around after normal hours. The boss is more accessible after hours than during the day, and the employee who is also around may pick up valuable contact time.

3. They get involved. Volunteer for more work, be a part of the solution, have opinions, and take stands. This commitment to getting involved in problem solving puts you on the firing line where shots can be taken at you. This is criticism.

4. They take criticism. Use criticism as feedback to make you better and stronger. Without criticism, we go on thinking that everything is OK. Do not be afraid of criti-

cism, or do not be overly embarrassed by criticism. A good attitude toward criticism will make life easier, and you will grow. To avoid criticism, do nothing, say nothing, and be nothing.

5. They criticize others. This is more difficult than being criticized, but if you think someone is saying or doing something wrong, it is your responsibility to say so. The proper attitude toward criticizing others is that if they persist, harm will come to them, and you want the best for them. Older, experienced people are able to criticize much too easily. One must never lose sight of the basic attitude that criticism is to help. Once a boss stops criticizing you, he or she has made a decision to remove you from the organization. Don't seek perfection; be reasonable and helpful.

6. They can be aggressive. There is no room for timidity in an aggressive organization. You must love the involvement and aggressively take an active part. Have opinions, share opinions, think outwardly, and have fun being creative. Truly, the only thing we have to fear is fear itself. Fear is not knowing and not being certain. So, if we learn all we can, we become certain and fear disappears.

I Will Not Wait on Others to Deliver the Goods

You cannot just ask for or order something and then sit back and wait. You must continually follow up until you get what you need. The squeaky wheel does get the grease. You can never blame others for not performing; only children do this.

I Will Confirm My Instructions by Asking Questions

Be sure you understand what is being asked of you. Ask, "Is this what you want?" and put it in your own words.

I Will Complete Everything I Start

Don't leave things half done. Some type of job list check-off system is needed to ensure completion. There is always an answer, so don't give up too soon.

I Will Say Nothing as Fact Unless I Can Back It Up

Opinion must be stated as opinion.

I Will Never Forget Who I'm Working For

The boss may not always be right, but he or she is always the boss and deserves your respect. You will never win by fighting with the boss. Who would want you working for them if you have a reputation of fighting the boss? Keep the boss informed and let him or her know that you know who the boss is. Sell your boss to everyone you meet. What the boss wants takes priority over what you want to do. The boss sets priorities.

Never go over or around the boss. No one will respect or admire your initiative. Getting the boss promoted is your best way of getting yourself promoted, and following the boss up the ladder can be very rewarding. Bosses sometimes go to other companies. If you've been a good supporter, they will ask you to join them.

Never Take Myself Too Seriously

A good sense of humor and a little self-depreciation will help your cause. Greet people with a smile, be friendly, and have some fun. A person is as happy as he or she wants to be, so learn to be happy. Think happy thoughts and you will be happy.

I Will Look and Act Professional

Clean, neat, business attire, polished shoes, and clean language are important to you and your career.

GOALS

To achieve great things, one must set great goals. To achieve your goals, you must first write them down and commit to them. If your goals are not written down, you have no goals. Each person must strive for something in his or her personal life and professional life, and personal goals must be in line with professional goals. In the same manner, professional goals must be aligned with company goals.

Goals must be reviewed periodically to check progress. Measurable goals are the best because progress can be measured. The review may also indicate a revision of goals. Everything changes, and goals are no exception. Time standards are goals just like a return on investment is a goal for an engineer and profitability is a goal for a company president.

Some Goals for the Motion and Time Study Department

1. To set fair and equitable time standards for the company and the employees. The number of grievances or complaints about time standards is a useful measurement of satisfaction, and a comparison of your plant's performance to industrial averages is a good measure. The image of your motion and time study department is at stake.

2. To have full time standards coverage. How many hours of work were performed on standard divided by total hours worked by a group is a good measure of coverage. The performance control system gave us a summary of all hours and classification of indirect work, and this total was divided by total hours worked to produce a percent indirect labor.

3. To promote productivity.

$$\% \text{ performance} = \frac{\text{earned hours}}{\text{actual hours}}$$

From our performance control system, discussed in chapter 12, the percent performance can be calculated for every person, department, shift, and plant total. Another useful technique of productivity measurement is units per manhour, dollars per manhour, pounds or tons per manhour, or board feet per manhour. These are usually old measures accepted industry-wide, and if they improve at the same rate as our percent performance, the time standards credibility improves.

4. To develop motion economy and cost consciousness in all employees, both labor and management. Each plant should have a cost reduction program, and each motion and time study person should save five times their annual salary. Developing the best possible methods is part of this.

5. To develop time standards for indirect labor. Set-up, material handling, quality control, rework, distribution, data processing, filing, etc. are all being studied and included in time standards systems. Expand the influences of motion and time study.

6. To develop standard data. Standard data is faster, more accurate, and easier to explain than any other method of setting time standards. The percentage of standards set using standard data is a good method of measuring standard data development and makes the department count the number of time standards set per period of time.

7. To teach employees and supervisors techniques of motion and time study. The number of people trained per year should be reported against a previously set goal.

8. To keep up with the developments in the field by attending conferences and subscribing to journals.

9. To prepare technologists for supervisory positions. What better place for training a supervisor?

10. To continually expand our knowledge of our company's technology and to become the expert in manufacturing methods, machines, techniques, and time standards.

11. To be ready to serve whenever asked and to be ready with the answer in our area of expertise.

QUESTIONS

1. Review the attitudes of an industrial technologist and discuss them until you are comfortable with them. If you understand them, you will make them a part of you.

2. List the goals of an industrial technologist.

3. Why must goals be written and measurable?

Forms

① Multiactivity Chart 2 sides
② Process Chart 2 sides
③ PTSS Form 2 sides
④ Time Study Form 2 sides
⑤ Rater Trainer Form 2 sides
⑥ Long Cycle Time Study Form 2 sides
⑦ Assembly Line Balance Form 2 sides

FRED MEYERS & ASSOCIATES

MULTI ACTIVITY CHART
☐ OPERATOR/MACHINE ☐ GANG ☐ MULTI MACHINE
☐ LEFT HAND/RIGHT HAND ☐ OPERATIONS

OPERATION NO.	PART NO.	OPERATION DESCRIPTION:
DATE:	TIME:	
BY I.E.:		

ACTIVITY	TIME IN MINUTES	ACTIVITY
	.05	
	.10	
	.15	
	.20	
	.25	
	.30	
	.35	
	.40	
	.45	
	.50	
	.55	
	.60	
	.65	
	.70	
	.75	
	.80	
	.85	
	.90	
	.95	
	1.00	
	1.05	
	1.10	

TOTAL UTILIZATION _____
% UTILIZATION _____

TOTAL UTILIZATION _____
% UTILIZATION _____

TIME	STUDY	CYCLE	COST:			
			HOURS PER UNIT _____	TOTAL NORMAL TIME IN MINUTES PER UNIT	
TOTAL			DOLLARS PER HOUR _____	+ ___ % ALLOWANCE	
OCC			DOLLARS PER UNIT _____	STANDARD TIME	
AVG. OCC				HOURS PER UNIT	
LEV FACT			LAYOUT & MOTION PATTERN ON NEXT PAGE	PIECES PER HOUR	
NORM. TIME					

·WORK STATION DESIGN·

LAYOUT   SCALE = _____

PRODUCT OR PART SKETCH   SCALE = _____

FRED MEYERS & ASSOCIATES   PROCESS CHART

☐ PRESENT METHOD   ☐ PROPOSED METHOD   DATE:_____   PAGE___OF___.

PART DESCRIPTION:

OPERATION DESCRIPTION:

SUMMARY	PRESENT		PROPOSED		DIFF.		ANALYSIS:		FLOW
	NO.	TIME	NO.	TIME	NO.	TIME			DIAGRAM
○ OPERATIONS							WHY	WHEN	ATTACHED
⇨ TRANSPORT.							WHAT	WHO	(IMPORTANT)
☐ INSPECTIONS							WHERE	HOW	
D DELAYS									
▽ STORAGES							STUDIED BY:		
DIST. TRAVELED		FT.		FT.		FT.			

STEP	DETAILS OF PROCESS	METHOD	OPERATION	TRANSPORT	INSPECTION	DELAY	STORAGE	DISTANCE IN FEET	QUANTITY	TIME HRS/UNIT .00001	COST PER UNIT	TIME/COST CALCULATIONS
1			○	⇨	☐	D	▽					
2			○	⇨	☐	D	▽					
3			○	⇨	☐	D	▽					
4			○	⇨	☐	D	▽					
5			○	⇨	☐	D	▽					
6			○	⇨	☐	D	▽					
7			○	⇨	☐	D	▽					
8			○	⇨	☐	D	▽					
9			○	⇨	☐	D	▽					
10			○	⇨	☐	D	▽					
11			○	⇨	☐	D	▽					
12			○	⇨	☐	D	▽					
13			○	⇨	☐	D	▽					
14			○	⇨	☐	D	▽					
15			○	⇨	☐	D	▽					
16			○	⇨	☐	D	▽					
17			○	⇨	☐	D	▽					

STEP	DETAILS OF (PRESENT/PROPOSED) METHOD	METHOD	OPERATION	TRANSPORT	INSPECTION	DELAY	STORAGE	DISTANCE IN FEET	QUANTITY	TIME .00001	COST PER UNIT	TIME/COST CALCULATIONS
18			○	⇨	☐	D	▽					
19			○	⇨	☐	D	▽					
20			○	⇨	☐	D	▽					
21			○	⇨	☐	D	▽					
22			○	⇨	☐	D	▽					
23			○	⇨	☐	D	▽					
24			○	⇨	☐	D	▽					
25			○	⇨	☐	D	▽					
26			○	⇨	☐	D	▽					
27			○	⇨	☐	D	▽					
28			○	⇨	☐	D	▽					
29			○	⇨	☐	D	▽					
30			○	⇨	☐	D	▽					
31			○	⇨	☐	D	▽					
32			○	⇨	☐	D	▽					
33			○	⇨	☐	D	▽					
34			○	⇨	☐	D	▽					
35			○	⇨	☐	D	▽					
36			○	⇨	☐	D	▽					
37			○	⇨	☐	D	▽					
38			○	⇨	☐	D	▽					
39			○	⇨	☐	D	▽					
40			○	⇨	☐	D	▽					
41			○	⇨	☐	D	▽					
42			○	⇨	☐	D	▽					

FRED MEYERS & ASSOCIATES PREDETERMINED TIME STANDARDS ANALYSIS

OPERATION NO.	PART NO.	OPERATION DESCRIPTION:
DATE:	TIME:	
BY I.E.:		

DESCRIPTION-LEFT HAND	FREQ.	LH	TIME	RH	FREQ.	DESCRIPTION-RIGHT HAND	ELEMENT TIME

TIME	STUDY	CYCLE	COST:				
			HOURS PER UNIT _____			TOTAL NORMAL TIME IN MINUTES PER UNIT	
						+ ___ % ALLOWANCE	
TOTAL			DOLLARS PER HOUR _____			STANDARD TIME	
OCC							
AVG. OCC			DOLLARS PER UNIT _____			HOURS PER UNIT	. _ _ _ _ _
LEV FACT						PIECES PER HOUR	
NORM. TIME							

LAYOUT   SCALE =

MOTION PATTERN

SEARCHING FOR A BETTER METHOD   ELIMINATE-COMBINE-CHANGE SEQUENCE-SIMPLIFY

FRED MEYERS & ASSOCIATES TIME STUDY WORKSHEET

☐ SNAP BACK
☐ CONTINUOUS

OPERATION DESCRIPTION

PART NUMBER	OPERATION NO.	DRAWING NO.	MACHINE NAME	MACHINE NUMBER
OPERATOR NAME	MONTHS ON JOB	DEPARTMENT	TOOL NUMBER	FEEDS & SPEEDS.

☐ QUALITY OK ?
☐ SAFETY CHECKED ?
☐ SETUP PROPER ?

NOTES:

PART DESCRIPTION:

MATERIAL SPECIFICATIONS:

MACHINE CYCLE TIME

ELEMENT #	ELEMENT DESCRIPTION		READINGS										TOTAL CYCLES	AVERAGE TIME	%	R	NORMAL TIME	FREQUENCY	UNIT NORMAL TIME	RANGE	R/X	HIGHEST
			1	2	3	4	5	6	7	8	9	10										
		R																				
		E																				
		R																				
		E																				
		R																				
		E																				
		R																				
		E																				
		R																				
		E																				
		R																				
		E																				
		R																				
		E																				
		R																				
		E																				

FOREIGN ELEMENTS:

NOTES:

R/X	# CYCLES
.1	2
.2	7
.3	15
.4	27
.5	42
.6	61
.7	83
.8	108
.9	138
1.0	169

TOTAL NORMAL MIN.
ALLOWANCE + _____ %
STANDARD MINUTES
HOURS PER UNIT
UNITS PER HOUR

ON BACK
WORK STATION LAYOUT
PRODUCT SKETCH

ENGINEER: _____ DATE: _ / _ / _

APPROVED BY: _____ DATE: _ / _ / _

WORK STATION DESIGN

LAYOUT   SCALE = _____

PRODUCT OR PART SKETCH   SCALE = _____

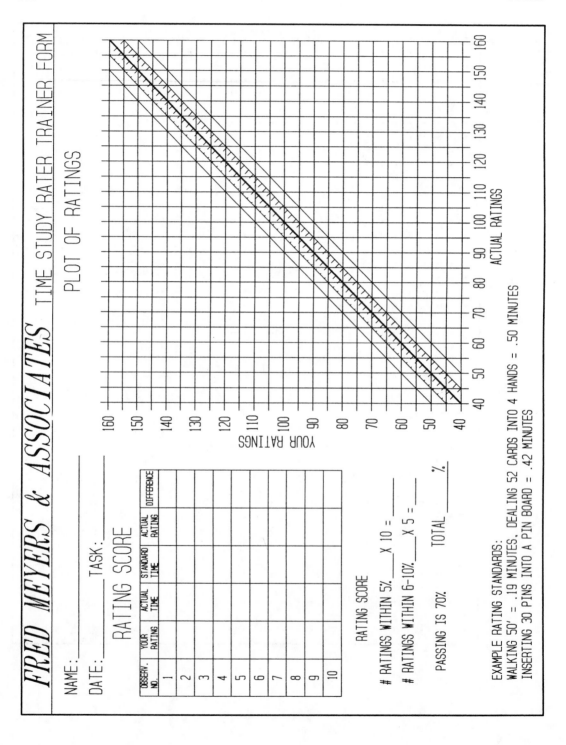

FRED MEYERS & ASSOCIATES TIME STUDY RATER TRAINER FORM

PLOT OF RATINGS

NAME: _____

DATE: _____ TASK: _____

RATING SCORE

OBSERV. NO.	YOUR RATING	ACTUAL TIME	STANDARD TIME	ACTUAL RATING	DIFFERENCE
1					
2					
3					
4					
5					
6					
7					
8					
9					
10					

RATING SCORE

RATINGS WITHIN 5% _____ X 10 = _____

RATINGS WITHIN 6-10% _____ X 5 = _____

PASSING IS 70%.   TOTAL _____ %.

EXAMPLE RATING STANDARDS:
WALKING 50' = .19 MINUTES. DEALING 52 CARDS INTO 4 HANDS = .50 MINUTES
INSERTING 30 PINS INTO A PIN BOARD = .42 MINUTES

TEN FUNDAMENTALS OF PACE RATING

1. Healthy people in the right frame of mind easily turn in 100% performance on correctly standardized jobs.
   For incentive pay, good performers usually work at paces from 115% to 135%, depending on jobs and individuals.

2. For most individuals *it is uncomforable to work at a tempo much below 100% and extrememly tiring to operate for sustained periods at paces lower than 75%;* our reflexes are naturally geared to move faster.

3. Poor efficiency on a correctly standardized job usually results from stopping work frequently—"goofing off" for a variety of reasons.
   Specifically, *substandard production seldom results from inability to work at a normal pace.*

4. Some standards of 100%
       1. Walking 3 m.p.h. or 264 ft./min.
       2. Dealing cards into four stacks in .5 min.
       3. Filling the pinboard in .435 min.

5. Very seldom can *true* performance of over 140% be found in industry.

6. Where an operator consistently comes up with extremely high efficiencies, it is usually a sign that the method has been changed or the standard was wrong in the first place.

7. An operator's work pace during a time study does not affect the final standard. His or her *actual* time is multiplied by the performance rule to give a job standard which is fair for all employees.

8. Inasmuch as healthy employees can *easily* vary work pace from approximately 80% to around 130%—through a range of 50%—reasonable inaccuracies in the setting of standards should be sensibly accepted.

9. Ineffective foremen usually fight job standards. *Good supervisors, however, sincerely help in the standard-setting effort, clearly realizing that such information is their best planning and control tool.*

10. Methods usually influence production more than work pace. Don't ever get absorbed in how quickly or slowly an operator "seems" to be moving that you fail to consider whether or not he or she is using the right method.

(Courtesy of Tampa Manufacturing Institute)

FRED MEYERS & ASSOCIATES LONG CYCLE TIME STUDY WORK SHEET

PART NO._____	OPERATION DESCRIPTION:
OPERATION NO._____	
DATE/TIME _____	MACHINE; TOOLS, JIGS:
BY I.E._____	MATERIAL:

ELEMENT #	ELEMENT DESCRIPTION	ENDING WATCH READING	ELEMENT TIME	% R	NORM. TIME

FRED MEYERS & ASSOCIATES MACHINERY & EQUIPMENT LAYOUT DATA SHEET

DESCRIPTION OF MACHINERY & EQUIPMENT: DATE:

DESCRIPTION OF OPERATION:

COMPANY NAME: LOCATION:

DESIGNED BY:

PHOTO

LAYOUT DRAWING

SCALE =

Reference Notations/Changes

MACHINE SPECIFICATIONS ON BACK.

FRED MEYERS & ASSOCIATES — ASSEMBLY LINE BALANCING

PRODUCT NO.: _____
DATE: _____
BY I.E.: _____

PRODUCT DESCRIPTION _____
NUMBER UNITS REQUIRED PER SHIFT _____

"R" VALUE CALCULATIONS

$$\text{EXISTING PRODUCT} = \frac{365 \text{ MINUTES}}{\text{UNITS REQ'D/SHIFT}} = \text{"R"}$$

$$\text{NEW PRODUCT} = \frac{300 \text{ MINUTES}}{\text{UNITS REQ'D/SHIFT}} = \text{"R"}$$

NO.	OPERATION/DESCRIPTION	"R" VALUE	CYCLE TIME	# STATIONS	AVG. CYCLE TIME	% LOAD	HRS./1000 LINE BALANCE	PCS./HR. LINE BALANCE

ASSEMBLY LINE LAYOUT   SCALE 1/4" = 1'

SEARCHING FOR A BETTER METHOD   ELIMINATE-COMBINE-CHANGE SEQUENCE-SIMPLIFY

Answer Sheet

CHAPTER 1

#1 page 1, line one;

#2 page 1, paragraph 2;

#3 page 2, paragraph 2—cost and quality;

#4 page 2, paragraph 1—we can reduce the cost of any job! . . . etc.;

#5 page 2, paragraph 3;

#6 page 2, paragraph 4—eliminate, combine, change sequence and simplify; eliminate is the most important;

#7 page 3, paragraph 1—6%, 85%, 120%;

#8 page 3, list—work hard, work long, criticize, and be criticized.

CHAPTER 2

Students should know what Taylor, Gilbreth, and Mayo did for motion and time study.

CHAPTER 3

#1 pages 17–28;

#2 page 15, paragraph 2;

#3 pages 16–17—decimal minute, pieces per hour and hours per piece;

#4 pages 24–25—60%, 85%, 120%;

#5 page 24—item 6;

#6 page 22:

   a) # machines: $480 - 48 = 432 \times 75\% = 324 \div 3000 = .108$
       $.284 \div .108 = 2.6$ machines—buy 3 machines

   b) cost: $60 \div .284 = 211$ pieces/hr.; $1/X = .00473 \times \$7.50 = \$.0355$

   c) quantity per shift: 211×8 hrs. $\times$ 3 machines $\times$ 75% performance $= 3798$ units per shift.

CHAPTER 4

#1 page 30, bottom;

#2 page 31, paragraph 2—PTSS;

#3 page 32, paragraph 8—stopwatch;

#4 page 34, bottom;

#5 page 37, paragraph 3—standard data;

#6 page 32, paragraph 6—PTSS.

CHAPTER 5

#1 page 44, paragraph 1;

#2 pages 45, 47, 51, and 56;

#3 pages 44 & 45—cost reduction;

#4 page 44—bottom—ask 6 questions of each step in the process to seek elimination, combination, reroute or simplification;

#5 pages 45 and 46;

#6 page 46, bottom;

#7 page 52, top;

#8 $.231 - .054 \times 1,000 = \$177,000.$

CHAPTER 6

#1 page 61, bottom;

#2 page 63, bottom;

#3 pages 62–64;

#4 page 59;

#5 page 67;

#6 page 67;

#7 comparison.

CHAPTER 7

#1 page 75–76;

#2 page 75, paragraph 3—cheapest cost, anything else must be justified;

#3 page 75, paragraph 3—anywhere;

#4 page 77, 78, 79, 80, 82, 83, 84, 86;

#5 page 86–87;

#6 page 86, bottom—blueprint of method and bill of material for standard.

CHAPTER 8

#1 pages 90–91, list of 9;

#2 page 91, middle;

#3 pages 92–97;

#4 page 100, paragraph 1;

#5 pages 96–97—design change or fixture;

#6 page 98, middle—eliminate;

#7 pages 102–103;

#8 page 104, middle—3 places on minutes, 5 places on hours;

#9 page 106, bottom;

#10 pages 106–109

PTSS Problem on page 109

R24	15	M24	.091
G3	9	RL	+.009
M24	15	R24	.100 minutes ÷ 60 =
1	1	G1	.00167 hours per minute 1/x =
M4	5	M4	
FM	6		600 units per hour
PROC	40		
	91 = .091 minutes		

CHAPTER 9

#1 see text;

#2 page 114, paragraph 1—oldest;

#3 page 114–115, bottom and top;

#4 page 116—list;

#5 pages 123–124;

#6 pages 126—list;

#7 pages 127–128;

#8 page 129;

#9 pages 129–130—list;

#10 page 130, middle;

#11 page 132, bottom;

#12 page 133, step 6;

#13 pages 140–143;

#14 page 151, bottom;

#15 page 139, step 9;

#16 pages 140–141, list;

#17 pages 143–147, 4 effort;

#18 page 142—79%, 119%, 68%, 173% 95%—page 143—125%, 111%, 91%, 83% and 136, 109, 87 and 97;

#19 page 147, middle;

#20 page 152;

#21 pages 152–154,

#22 pages 155–157;

#23 page 158;

#24 pages 158–159.

CHAPTER 10

#1 page 166, paragraph 1;

#2 pages 166–167 (1–8);

#3 page 167 (1–5)

#5 page 174: $.052 + .55 + .24 + .125 + .050 + .110 = 1.397 \times 110\% = 1.537 \div 60 = .0256$ hrs.; $1/x = 39$ per hour;

#6 page 175:

(1) $\dfrac{250 \times 12}{3.14 \times .5} = 1911$ RPM   (2) $\dfrac{3}{.001} = 3000$ rev.   (3) $\dfrac{3000 \text{ REV}}{1911 \text{ RPM}} = 1.57$ min.

#7 Drill—add .4 times the diameter (.5) to the length of cut (3.0″). Mill—add the diameter of the cutter to the length of cut and multiply the number of teeth on the cutter times the feed rate .001.

CHAPTER 11

#1 page 180, paragraph 1;

#2 page 180, Elemental Ratio Studies, Performance Sampling Studies and Time Standard Development Studies. I—page 188, II—III—page 189;

#3 page 180–181;

#4 a, b, c, d, e, f, g all on pages 181–184;

#5 pages 185–188;

#6 pages 188–189;

#7

Job	No. Observations	%	Hours	No. Units	Hrs./Unit Normal	Standard Hrs./Unit
1	5000	10	425	2900	.14655	.16121
2	10,000	20	850	8800	.09659	.10625
3	20,000	40	1700	25,000	.06800	.0748
Idle	15,000	30	1275			
Total	50,000	100	4250			

70% + 7% = 77% Efficiency

CHAPTER 12

#1 page 194;

#2 page 194, list;

#3 page 195, 85%;

#4 85% of 85% = 72.25%;

#5 page 197;

#6 page 197;

#7 a, b, c page 197–198;

#8 page 199–200

#9

Act. Hrs.	Earned Hrs.	%
2	2.40	120
1.75	1.67	95
2.75	2.67	97
1.5	2.00	133
8	8.74	109

#10 pages 203–204.

CHAPTER 13

#1 pages 210–211;

#2 a, b, c pages 210–211;

#3 page 212;

#4 page 213;

#5 page 214;

#6 individual incentive plans, group plans;

#7 page 215;

#8 26%;

#9 page 216;

#10 pages 216–217;

#11 page 217;

#12 pages 217–218;

#13 pages 219–223;

#14 pages 223–224;

#15 page 224.

CHAPTER 14

No questions.

CHAPTER 15

#1 page 256;

#2 page 256;

#3 page 257;

#4 page 260;

#5 page 260;

#6 page 261;

#7 page 260;

#8 page 262;

#9 page 262;

#10 any improvement with a total hours per 1000 of less than 43.051 is a better layout;

#11 pages 262–263;

#12 49.36, 4.54, .45;

#13 page 265 is an improvement of page 264.

CHAPTER 16

#1 page 268;

#2 page 268;

#3 page 269;

#4 pages 269–276;

#5 You are on your own. Use your knowledge of each technique.

CHAPTER 17

#1 pages 278–279, list;

#2 on your own;

#3 on your own.

CHAPTER 18

#1 on your own;

#2 page 289;

#3 page 289.

Index